本书受到国家自然科学基金支持：

基于微观结构特征的蒙药"孟根乌苏"物质基础及其免疫分子调节机制研究，

项目编号：81503351

蒙药黑冰片的药效物质基础及其对胃寒性"协日"病的作用机制研究，

项目编号：81760750

中空/多孔纳米结构的可控制备及其性能研究

武世奎 著

燕山大学出版社

·秦皇岛·

图书在版编目（CIP）数据

中空/多孔纳米结构的可控制备及其性能研究 / 武世奎著. —秦皇岛：燕山大学出版社，2023.3

ISBN 978-7-81142-186-6

I. ①中… II. ①武… III. 纳米材料－材料制备－研究②纳米材料－性能－研究 IV. ①TB383

中国版本图书馆 CIP 数据核字（2020）第 082213 号

中空/多孔纳米结构的可控制备及其性能研究
武世奎 著

出 版 人：陈 玉	
责任编辑：朱红波	策划编辑：朱红波
责任印制：吴 波	封面设计：刘韦希
出版发行： 燕山大学出版社 YANSHAN UNIVERSITY PRESS	电 话：0335-8387555
地 址：河北省秦皇岛市河北大街西段 438 号	邮政编码：066004
印 刷：秦皇岛墨缘彩印有限公司	经 销：全国新华书店
开 本：700mm×1000mm 1/16	印 张：10
版 次：2023 年 3 月第 1 版	印 次：2023 年 3 月第 1 次印刷
书 号：ISBN 978-7-81142-186-6	字 数：156 千字
定 价：40.00 元	

版权所有 侵权必究

如发生印刷、装订质量问题，读者可与出版社联系调换

联系电话：0335-8387718

前　言

中空/多孔纳米结构材料具有合适的空腔/孔隙大小、较大的比表面积和增强的功能性质，这使其较实心纳米颗粒具有更广泛的应用前景。由于结构相对复杂，目前制备具有中空/多孔结构的功能纳米复合物仍是一项具有挑战性的工作。发展可设计的制备方法是实现中空/多孔结构功能复合物广泛应用的关键。本书从"自下而上"和"自上而下"两种途径来探索这类纳米结构的可控制备方法。本书第一部分内容（第2章、第3章）采用"自下而上"途径，选择自模板法在液相中制备了AgCl中空立方块，进而通过光照在AgCl中空立方块表面原位生长了银纳米颗粒，形成Ag/AgCl复合光催化剂。本书第二部分内容（第4～7章）采用"自上而下"途径，尝试用一种简单的金属-有机框架（MOF）为模板制备中空/多孔纳米材料。首先制备了几种不同形貌的普鲁士蓝类似物（PBA）的纳米颗粒，包括多孔结构$Zn_3[Fe(CN)_6]_2·xH_2O$和$NiFe(CN)_5NO·2H_2O$。再以PBA纳米颗粒作为模板，通过高温热解的方法"自上而下"制备了组分更加丰富的中空/多孔纳米结构复合物。由于其结构和性能更易调控，这些中空/多孔纳米复合材料在吸附、光催化、电催化析氢（HER）等多个领域得到了综合应用。主要研究内容如下：

1. 以$AgNO_3$和CCl_4为原料，通过简单的一步法制备了方块状的AgCl中空纳米结构。该AgCl中空立方块的边长为600～900 nm，孔腔尺寸400～600 nm。溶解-沉淀与离子扩散相结合的形成机理用于解释中空纳米结构的生长过程。通过光辐射可以将AgCl中空方块转化为Ag/AgCl中空方块。该复合结构表现出卓越的有机污染物降解活性。通过对上述实验的改进，以$AgNO_3$和CCl_4作为前驱体、PVP为保护剂，利用溶剂热法制备了大量均匀的尺寸100 nm以下的AgCl纳米方块和纳米球。研究表明，反应时间和温度是产品形貌演变的重要因素。如果以二氯乙烷作为氯源,则可获得四面体型的 AgCl 纳米结构。样品的形貌演化规律表明其生长机理可能遵循奥斯特瓦尔德机制。该结构同样可转化成 Ag/AgCl 复合光催化剂,并在太阳光下展现出优异的罗丹明 B 降解效率。

2. 以 $ZnCl_2$ 和 $K_3Fe(CN)_6$ 为原料,成功制备了规则的 $Zn_3[Fe(CN)_6]_2 \cdot xH_2O$ 球状纳米颗粒、微米立方块和多面体状颗粒。通过调节反应体系中盐酸的用量,实现了 $Zn_3[Fe(CN)_6]_2 \cdot xH_2O$ 形貌从立方块向多面体转化。通过 X 射线衍射(XRD)、扫描电子显微镜(SEM)、透射电子显微镜(TEM)和比表面积分析对样品进行了表征,并提出了各种形貌颗粒的可能形成机理。$Zn_3[Fe(CN)_6]_2 \cdot xH_2O$ 具有沸石样结构,对有机染料有较强的选择性吸附作用。对染料吸附的实验结果显示,立方块状样品对亚甲基蓝吸附能力最强,达到 1.016 $g \cdot g^{-1}$(比表面积达到 643.2 $m^2 \cdot g^{-1}$)。并且吸附的染料可以在有机溶剂中脱附,实现染料的回收利用。因此,这一材料可用于废水处理领域。

3. 基于 MOF 的多孔结构,$Zn_3[Fe(CN)_6]_2 \cdot xH_2O$ 可用于吸附重金属离子 Ag^+ 形成 Ag 掺杂的 $Zn_3[Fe(CN)_6]_2 \cdot xH_2O$(Ag-ZnPBA)。通过煅烧块状的 Ag-ZnPBA 可获得均匀的具有中空或多孔结构的三元 $Ag/ZnO/ZnFe_2O_4$ 复合物纳米材料。样品在煅烧后保持了前驱体的块状形貌;并且通过调节加热程序,三元复合物的微观结构可以实现从多孔到中空的转化;银的负载量可以通过前驱体对银离子的吸附量加以调节。银负载的三元复合物 $Ag/ZnO/ZnFe_2O_4$ 与其二元复合物 $ZnO/ZnFe_2O_4$ 相比表现出更强的光催化活性,说明银在光催化过程中起到了重要作用。结果表明复合物中的 Ag、ZnO 和 $ZnFe_2O_4$ 三种组分形成了 II 型匹配和肖特基势垒,从而促使光生电子和空穴更好地分离。同时由于铁酸锌卓越的磁性,可以方便地用磁铁将光催化剂从反应体系中分离出来。该研究提供了一种制备均匀的多功能多组分中空结构的简单有效的途径。

4. 在氮气保护下煅烧 $Zn_3[Fe(CN)_6]_2 \cdot xH_2O$ 多面体微块,颗粒可以保持多面体形貌，并在多面体颗粒表面长出了中空的碳纳米管。XRD、电子能谱、X 射线光电子能谱结果显示其组分为 Fe、Fe_3C 和氮掺杂碳。高分辨透射电子显微镜（HRTEM）结果显示，多面体颗粒内部是由 10～50 nm 的 Fe/Fe_3C @N-doped C 核壳结构纳米颗粒堆积而成。Fe/Fe_3C@N-doped C 纳米颗粒之间堆积比较疏松，形成多孔结构。这种由氮掺杂碳包覆过渡金属的纳米材料具有良好的 HER 性能。电化学测试表明，该催化剂的过电位和 Tafel 斜率较小，分别为 109.1 mV 和 59.6 mV dec^{-1}，是良好的 HER 电催化剂。磁性测试结果表明，该材料具有较高的饱和磁化强度和矫顽力，在存储材料领域具有应用潜力。

5. 在氮气保护下高温煅烧尺寸约 50 nm 的 $NiFe(CN)_5NO·2H_2O$ 纳米颗粒，我们获得了 NiFe@氮掺杂石墨烯管状结构（NiFe@NGT）。XRD、SEM 和 TEM 结果显示，该结构由直径为 1～2 μm 的氮掺杂石墨烯管和封装在管中的 NiFe 合金颗粒组成。通过调节煅烧温度，石墨烯管的直径可调节到 50～200 nm。结构和组分表征结果表明，前驱体在氮气下高温煅烧时会发生熔合，形成微米级的 NiFe 合金颗粒，并催化形成氮掺杂石墨烯微管。该结构可为 HER 提供更多的活性位点。电化学测试表明，该催化剂的过电位和 Tafel 斜率较小，分别为 70.5 mV 和 63.4 mV dec^{-1}，且循环性能稳定，是优异的 HER 催化剂。

目 录

第1章 绪论 1
  1.1 引言 1
  1.2 中空纳米结构的液相法制备 1
      1.2.1 硬模板法 1
      1.2.2 软模板法 3
      1.2.3 无模板法 4
  1.3 MOF 前驱体法制备中空/多孔纳米结构 6
      1.3.1 MOF 纳米颗粒的可控制备及吸附性能 6
      1.3.2 在空气中煅烧 MOF 纳米颗粒 7
      1.3.3 在惰性气体保护下煅烧 MOF 纳米颗粒 9
  1.4 中空/多孔纳米结构的应用 11
      1.4.1 光催化 11
      1.4.2 电化学析氢反应（HER） 17
  1.5 本书研究的内容 20

第2章 中空 Ag/AgCl 块状纳米结构的制备及其光催化性能研究 21
  2.1 引言 21
  2.2 材料和方法 21
      2.2.1 材料 21
      2.2.2 AgCl 中空方块的制备 22
      2.2.3 仪器 22
      2.2.4 光催化性能 22
  2.3 结果与讨论 23
      2.3.1 结构与形貌 23

2.3.2 AgCl 中空方块的形成过程 24
2.3.3 AgCl 中空方块的生长机理 26
2.3.4 中空 Ag/AgCl 方块的光催化性能 27
2.4 本章小结 32

第 3 章 Ag/AgCl 纳米结构的形貌可控制备及其光催化性能研究 33

3.1 引言 33
3.2 材料和方法 33
    3.2.1 材料 33
    3.2.2 制备方法 34
    3.2.3 仪器 34
    3.2.4 光催化性能 34
3.3 结果与讨论 35
    3.3.1 AgCl 纳米晶的形貌可控制备 35
    3.3.2 反应温度和时间对形貌的影响 36
    3.3.3 反应溶剂对形貌的影响 39
    3.3.4 Ag/AgCl 纳米晶的光催化性能 44
3.4 本章小结 47

第 4 章 多孔六氰合铁酸锌（$Zn_3[Fe(CN)_6]_2 \cdot xH_2O$）微-纳米晶的制备及吸附性能研究 48

4.1 引言 48
4.2 实验部分 49
    4.2.1 材料 49
    4.2.2 制备 ZnPBA 微-纳米晶 49
    4.2.3 仪器 49
    4.2.4 染料吸附实验 50
4.3 结果与讨论 50

        4.3.1 制备与表征 50
        4.3.2 反应条件的影响 52
        4.3.3 ZnPBA 微-纳米晶的吸附性能 55
    4.4 本章小结 61
第 5 章 多孔/中空 Ag/ZnO/ZnFe$_2$O$_4$ 三元复合物的制备及其光催化性能
        研究 62
    5.1 引言 62
    5.2 实验部分 63
        5.2.1 材料 63
        5.2.2 材料的制备 63
        5.2.3 仪器 64
        5.2.4 光催化性能 64
        5.2.5 光电性能 65
    5.3 结果与讨论 65
        5.3.1 制备与表征 65
        5.3.2 光催化性能 72
    5.4 本章小结 79
第 6 章 分级结构的 Fe/Fe$_3$C@N doped C 复合物的制备及其 HER 性能
        研究 80
    6.1 引言 80
    6.2 实验部分 81
        6.2.1 材料 81
        6.2.2 分级核壳结构 Fe/Fe$_3$C@N doped C 纳米复合物的制备 81
        6.2.3 仪器 81
        6.2.4 电化学性能测试 82
    6.3 结果与讨论 82

6.3.1 结构和形貌表征 82
6.3.2 分级结构的 Fe/Fe$_3$C@N doped C 复合物的形成机理 88
6.3.3 分级结构的 Fe/Fe$_3$C@N doped C 复合物的 HER 性能 88
6.4 本章小结 90

第 7 章 NiFe@N 掺杂石墨烯微管的一步法制备及其 HER 性能研究 91

7.1 引言 91
7.2 实验部分 92
    7.2.1 材料 92
    7.2.2 氮掺杂石墨烯管复合物的制备 92
    7.2.3 仪器 92
    7.2.4 电化学性能测试 93
7.3 结果与讨论 93
    7.3.1 氮掺杂石墨烯管的表征 93
    7.3.2 NiFe-NGT 复合物的形貌转化 100
    7.3.3 NiFe-NGT 复合物的 HER 性能 102
7.4 本章小结 105

参考文献 106

第1章 绪　　论

1.1 引言

由于具有独特的结构、规则的形貌、低密度、高表面积等特征，中空/多孔纳米结构的可控制备在过去几十年里一直备受关注[1]，其在光电、磁学、催化等领域有着广泛的应用[2-4]。本书尝试以液相法和金属有机框架（MOF）前驱体法构筑中空/多孔的复合纳米结构，研究中空/多孔纳米结构的生长机理和相关性能，旨在探索具有普遍意义的制备方法和实际应用。本章拟从液相法制备中空纳米结构，MOF 煅烧法制备中空/多孔结构功能复合物，以及其光催化、电催化性能三个方面综述其研究进展。

1.2 中空纳米结构的液相法制备

中空纳米结构的液相法制备主要包括：硬模板法、软模板法和无模板法。

1.2.1 硬模板法

硬模板法采用聚合物、无机非金属或金属颗粒作为模板。通过反应或表面作用在模板包裹壳层物质，再采用煅烧或有机溶剂溶解等方法去除模板，最终获得中空/多孔纳米结构[5-9]。硬模板法最终获得产品的形貌和粒径依赖于模板，且产品的形貌和单分散性好。

聚合物胶体是制备中空结构时最常用的模板。通常聚苯乙烯及其衍生物胶体颗粒被作为模板用于制备 SiO_2、SnO_2、CoO、Fe_2O_3 等[10-13]。在制备过程中，通常在溶液中将无机分子前驱体通过表面沉积或者表面反应均匀地覆盖到聚苯乙烯模板上，形成核壳结构的复合物。随后聚乙烯模板可以通过溶解或者煅烧的方式除去。Yang 等人采用一种磺化聚乙烯颗粒的过程制备了中空纳米颗粒[14],同时他们还获得了双层结构的 TiO_2 中空球[15]、C[16]、$BaTiO_3$ 和 $SrTiO_3$ 的中空颗粒[17]。Wu 等报道了一步法制备单分散的 SiO_2、TiO_2 和 ZnO 中空球的过程[18-20]（见图 1.1）。他们将壳的沉积和聚合物核的溶解过程在同一溶剂中进行。

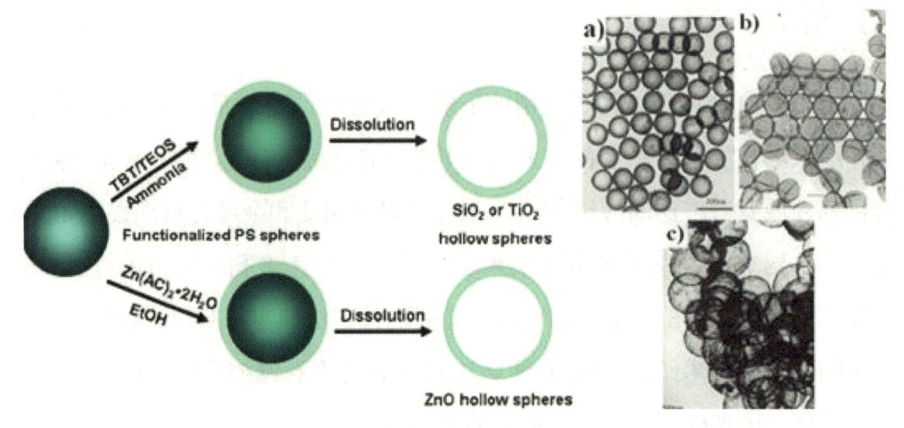

图 1.1　中空球制备过程的机理图及（a）SiO_2、（b）TiO_2 和（c）ZnO 中空球的 TEM 图

无机非金属模板主要为碳和硅的纳米颗粒。碳纳米颗粒表面有许多可反应的基团使壳材料更容易沉积，作为模板其更容易通过加热除去。Yang 等人[21] 报道了以碳纳米颗粒为模板制备的 Gd_2O_3 中空微球，其直径约 300 nm。Xia 等[22] 以介孔碳球为模板，采用同样的方法制备了几种多孔晶态氧化物中空球（TiO_2、ZrO_2、Al_2O_3 和 MgO 等）。Li 等人通过在亲水性的碳纳米颗粒表面吸附金属阳离子的方法，制备了尺寸可控的 Ga_2O_3 和 GaN 中空微球[23]。利用这一技术，同样可以获得 $CoFe_2O_4$、Pd/SiO_2 等的中空/多孔纳米颗粒[24-25]。

硅作为模板更容易控制和调节尺寸。硅作为模板通常先让合适的无机前驱体在硅纳米颗粒表面沉积，而后用碱液或氟化氢溶液去除硅纳米颗粒，形成中空纳米结构。通过这一方法可以获得 C、$CuSiO_4$、ZnS 等[26-28]。

与上述模板不同的是，金属模板本身会作为反应物参与反应。而该法制备的中空结构的形貌、孔洞、壁厚直接由金属模板决定。通常金属模板法制备中空结构的过程分为两种：柯肯达尔效应和置换反应。

柯肯达尔效应原来是指两种扩散速率不同的金属在热扩散过程中会形成缺陷。在纳米尺度内，以金属颗粒作为模板通过柯肯达尔效应可以获得中空结构的纳米颗粒[29]。Alivisatos 等首次用柯肯达尔效应解释了钴的氧化物和硫族化合物纳米结构的制备过程。在形貌转化过程中，钴原子由内向外壳扩散，在颗粒内部形成空腔，同时壳层的钴与溶液中的硒反应生成 CoSe 的外壳[30]。内部的金属钴与壳层的 CoSe 之间形成不连续的空腔，金属钴继续反应直到消耗完毕。由于最终产物的空腔比原始 Co 模板的直径小，说明阴离子也向内扩散，但其扩散速率较阳离子慢。通过进一步的研究，该课题组也获得了 CdS[31] 和 Fe_2O_3[32] 中空纳米颗粒，并发现扩散平衡，阴离子的浓度和反应活性对中空结构的形成有重要影响[33]。

通过柯肯达尔效应获得的纳米颗粒往往具有多晶结构，只有少数具有单晶结构[34]。尽管通过柯肯达尔效应可以制备一些中空纳米颗粒，但反应条件特殊，有一定的局限性。与聚合物和非金属模板相比，柯肯达尔效应制备的中空颗粒的结构边缘比较模糊。柯肯达尔效应还只是从现象学和理想模型角度分析其可能性，确切的机制仍需深入研究。

置换反应机制提供了一个用于金属中空纳米结构可控合成的非常简单的途径。这一过程的关键在于，金属模板的活性要比溶液中作为前驱体的金属离子要高，且金属离子浓度较低。Xia 等通过置换银纳米颗粒的方法获得了 Au、Pt、Pd 的中空纳米颗粒[35]。与柯肯达尔效应比较的话，原电池置换法不需要功能化表面，并且可以作为制备金属中空颗粒的通用方法，但也只能用于金属中空颗粒的制备。

1.2.2 软模板法

软模板法是用大量分子形成的结构相对稳定的分子体系，如生物大分子、微乳液、泡囊等作为模板制备中空/多孔纳米结构的方法。软模板比较容易去除，制备中空纳米颗粒更具优势[36-37]。然而，由于软模板容易变形，所制备的中空纳米颗粒的单分散性和形貌规则性较差。因此，控制无机中空结构的单分散性和规则形貌是该技术最具挑战性的部分。

Imhof 等利用软模板法成功制备了微米尺度的 SiO_2 的中空球。该模板为低分子量的聚二甲基硅氧烷（PDMS）有机硅的 O/W 乳状液液滴，直径在 0.6～2 μm 之间[38]。Han 等利用水/正庚烷/CTAB 的微乳液制备了 SiO_2 的中空球[39]。但是由于微乳液液滴的不稳定性，其容易受到加热、pH、离子强度的影响。

1.2.3 无模板法

无模板法是一种在液相合成中通过 Ostwald 熟化过程形成空腔结构的中空/多孔纳米材料制备方法。在无模板中空/多孔纳米结构时，反应物先成核，在表面能最小化的驱动下，大小不同的亚稳态的晶核形成聚集体，聚集体中的颗粒彼此连接，且存在空隙。Ostwald 熟化过程中由于小颗粒表面的母液浓度比大颗粒高，导致小颗粒溶解、大颗粒逐渐生长。这会使较大聚集体的壳层变厚，而小的聚集体消失。而颗粒内部溶质也会通过渗透作用向壳层扩散，导致内部空腔变大，最终形成中空结构[40]。

利用 Ostwald 熟化机制已经制备了大量的中空纳米颗粒，如 Fe_3O_4、Co、Sb_2S_3、Bi_2WO_6、TiO_2 等[41-45]。Wang 等以 β-$Ni(OH)_2$ 为原料制备了分级结构的 $Ni(OH)_2$ 中空微球[46]。通过在 600 ℃下煅烧 $Ni(OH)_2$ 中空微球 2 h 还可以得到 NiO 的中空球。$Ni(OH)_2$ 中空微球的形成过程可分为三步：首先沉淀出的 $Ni(OH)_2$ 组装成松散的聚集体，直径约 800～900 nm。随着反应的进行，聚集体的尺寸和密度增加形成实心球。最后通过类似 Ostwald 熟化的过程，内核的向外疏散过程逐渐形成空腔[47]。

在一种光催化活性高的介孔锐钛矿相 TiO_2 中空球的制备过程中，采用表面光滑、非晶态的 TiO_2 实心球作为前驱体。经过 30 min 水热反应后，产品的 XRD 图谱显示出现了结晶 TiO_2 锐钛矿相的衍射峰。TEM 结果显示产品仍然是实心球,球体的表面变得粗糙,且含有晶态的纳米颗粒。延长反应时间至 60 min,样品的结晶度增加,颗粒内部出现了镂空结构[48]。这一方法也可看作以自身作为模板的"自模板法"。

通过 Ostwald 熟化过程还可以制备更加复杂的纳米结构,如核壳结构、二元和三元复合物中空结构[49]。Zeng 等用该技术首先合成了氧化物和硫化物半导体的中空核壳结构[50]。微晶聚集体的原始形貌和聚集方式决定了 Ostwald 熟化后颗粒的最终形貌。随后他们证明了 Ostwald 熟化还可用于制备掺杂的中空纳米结构。通过 TiF_4 和 SnF_4 水解可制备出高质量的 Sn 掺杂 TiO_2 中空球(见图1.2)[51]。

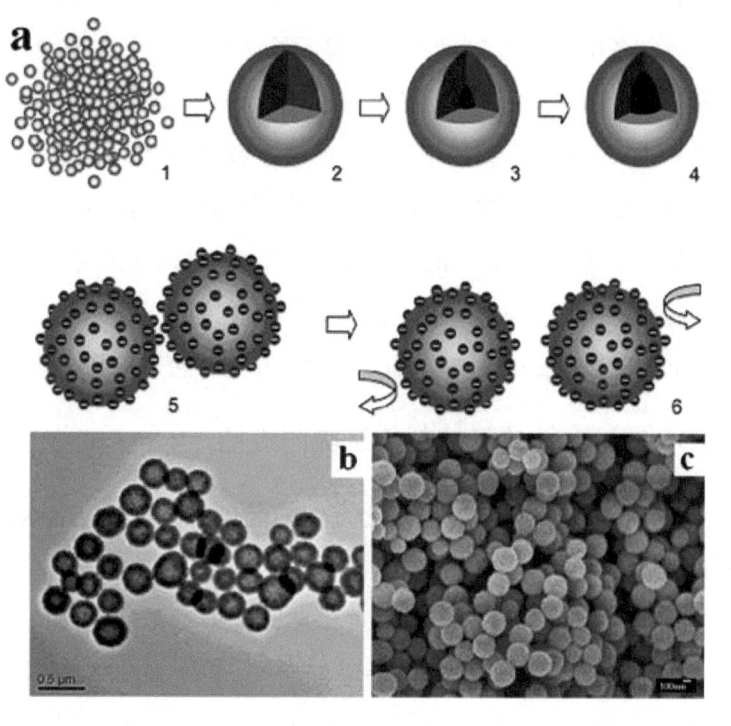

图 1.2 (a)Sn 掺杂的 TiO_2 中空球的形成过程及其(b)SEM、(c)TEM 图片[51]

尽管无模板法克服了上述模板法的问题,但是这一方法的应用只针对少数化合物,且反应机制不明确。对无机中空球的 Ostwald 熟化制备机理研究需要提供更深入的基础研究。因此本书通过研究 AgCl 中空纳米块的制备，对无模板法制备中空纳米颗粒的过程提出了新的理论。但液相法在制备过程中需要特殊的反应体系和大量的实验探索，导致在设计、制备功能复合物纳米结构时遇到了较大的困难[52]。

1.3 MOF 前驱体法制备中空/多孔纳米结构

近年来，研究人员注意到 MOF 作为前驱体通过化学或物理方法处理后可以直接获得具有独特结构和性能的纳米复合物且功能可调。经过几十年的研究，MOF 已形成一个庞大的家族，其纳米结构的制备同样受到了极大的关注。通过煅烧 MOF 纳米前驱体的方法可以获得中空/多孔纳米结构。MOF 作为前驱体制备中空/多孔纳米结构具有明显的优势：（1）MOF 纳米材料通常具有多金属组分，煅烧后的复合材料继承了前驱体的多组分特征，因此该法很容易获得多元复合物，且各组分严格遵循化学计量关系。（2）MOF 材料在煅烧过程中有机配体分解时释放气体，使得煅烧后的复合材料更容易获得多孔的中空结构。（3）MOF 纳米颗粒的形貌可以通过简单条件变化来调节，而煅烧后的复合材料会继承前驱体的形貌特征。因此，以 MOF 为前驱体"自上而下"制备中空/多孔纳米结构，其组分和形貌的可设计性更好。

1.3.1 MOF 纳米颗粒的可控制备及吸附性能

从 2000 年 Mann 等制备普鲁士蓝纳米晶以来，可控制备 MOF 纳米结构已成为纳米材料研究的热点之一[53]。经过十余年的努力，研究人员发展了各种用于 MOF 纳米颗粒形貌和尺寸可控制备的方法，并且提出了其成核与生长理论[54]。目前各种维度的 MOF 纳米结构均有制备[55]，并且更新颖的结构不断被报道，如核壳结构[56]、空心结构[57]、异质外延结构[58] 等。这使得 MOF 纳米晶作为多功能中空纳米复合物的模板在形貌和组分上具有了优势。

Nune 等系统研究了 MOF 纳米材料的吸附动力学,他们认为纳米尺度的 MOF 材料较其大块材料有更好的吸附性能是由于减少了扩散路径并优化了晶面相互作用[59]。Azadbakht 等制备了一种 Ni-MOF 纳米颗粒并对其碘离子吸附性能进行了研究[60]。Shooto 等制备了一种新颖的 MOF 纳米纤维并测试了其对铅离子的吸附性能[61]。Huang 等制备了一种高度稳定的 MOF 纳米材料,该材料可以将苯与其他芳香类物质有效分离[62]。基于 MOF 材料的多孔性质,其纳米材料被广泛用于吸附水中的有机染料、重金属离子、油污等,在水污染治理领域具有较好的应用前景。此外,利用 MOF 的这一性质,可以通过定量吸附的方法增加其组分,在煅烧过程中有望获得组分更加丰富的中空/多孔纳米颗粒。

1.3.2 在空气中煅烧 MOF 纳米颗粒

在空气中煅烧普鲁士蓝(PB)可以获得具有分级结构的中空 Fe_2O_3 微米盒子。在煅烧过程中 PB 分解,Fe_2O_3 的晶体原位生长,通过调节煅烧温度,可以获得不同的壳层结构。首先制备高度均匀的 PB 微块。接下来的加热过程可分为三个阶段(见图 1.3):Ⅰ阶段,在 350 ℃ 以下加热,使颗粒表面的 PB 分解为 Fe_2O_3。在持续的加热过程中,PB 颗粒的内部氧化分解产生的气流向外扩散,导致颗粒表面形成 Fe_2O_3 外壳,内部形成大的空腔。Ⅱ阶段,提高煅烧温度至 550 ℃,由于 Fe_2O_3 晶体的原位生长,Fe_2O_3 颗粒逐渐长大,颗粒表面的壳转化为多孔结构。Ⅲ阶段,煅烧温度增加到 650 ℃,壳上的 Fe_2O_3 的多孔结构继续生长形成分级结构。Yamauchi 等研究表明小颗粒的 PB 前驱体煅烧后容易形成多孔 Fe_2O_3,大颗粒 PB 煅烧后形成中空 Fe_2O_3。并且通过调节煅烧温度还可以形成纯净的 $\alpha\text{-}Fe_2O_3$ 或 $\beta\text{-}Fe_2O_3$。这一结果表明,除了继承前驱体的形貌,煅烧过程的晶体原位生长会产生更加新颖的二级结构,这将丰富中空纳米结构的性能,且拓展其应用领域[63-65]。

普鲁士蓝类似物(PBA)是一类简单的具有立方晶格结构 MOF 材料,

通式为 $M_3^{II}[M^{III}(CN)_6]_2 \cdot nH_2O$。以 $M_3^{II}[Co(CN)_6]_2 \cdot nH_2O$（M = Co，Mn，Fe，Zn，Cd）为例，通过改变 M 离子可获得不同组分的 Co-PBA 纳米颗粒。如果反应在不同浓度或种类的表面活性剂的辅助下完成，则可获得不同形貌的纳米颗粒，如球形、块状、八面体状等。再通过煅烧反应，可获得一元或二元氧化物的多孔/中空纳米结构[66-68]。

图 1.3 分别在（a~c）350 ℃，（d~f）550 ℃ 和（g~i）650 ℃ 制备的中空结构 Fe_2O_3 立方块的 SEM 和 TEM 图片[63]

用类似的方法，Ogale 等利用 MOF-199 制备了 CuO 的多孔纳米结构[69]。而 Hu 等煅烧 $[Cu_3(btc)_2]_n$ 前驱体，获得了 CuO/Cu_2O 的中空纳米结构[70]。Yan 等煅烧 $Sn_m[Fe(CN)_6]_n$ 获得了多孔的铁掺杂 SnO_2 纳米结构，该材料具有优异的锂离子存储性能[71]。Hou 等用 Bi 离子掺杂的 $Zn_3[Fe(CN)_6]_2$ 作为前驱体在氮气下煅烧获得了多孔的块状 $Bi/ZnO/ZnFe_2O_4$ 纳米结构。该方法为下一代锂离子电池阳极材料的可设计制备提供了新的可能[72]。

1.3.3 在惰性气体保护下煅烧 MOF 纳米颗粒

Salunkhe 等利用 Zn 基 MOF 在不添加任何其他物质的情况下制备了纳米多孔碳材料。他们采用的方法是在氮气气氛中以 5 ℃·min^{-1} 的升温速率至 800 ℃ 加热 5 h。之后将获得的粉末在 10% HF 溶液中浸泡 24 h，用于除去 Zn 纳米颗粒。该多孔碳材料可作为优异的超级电容器电极材料[73]。Lou 的课题组煅烧 NENU-5 获得了八面体形的 MoC$_x$ 的中空多孔纳米结构。他们首先利用简单的方法获得了 NENU-5 八面体纳米结构，进而以 2 ℃·min^{-1} 的升温速率至 800 ℃ 并保持 6 h，获得 MoC$_x$-Cu 中空纳米结构。最后他们用 0.1 M 的 FeCl$_3$ 溶液将 Cu 腐蚀掉，获得了八面体形的 MoC$_x$ 的中空多孔纳米结构。该结构具有优异的电化学析氢（HER）性能[74]。Lee 等在 ZIF-8 的表面覆盖了一层聚苯乙烯，并将其在氮气下加热到 1000 ℃，从而获得了分散性好的中空多孔碳球。该结构具有良好的吸附性能[75]。

将 MOF 原位生长于碳材料（如碳纳米管、石墨烯等）上，并通过进一步的煅烧可以获得性能优异的电极材料。Ma 等将 Zn$_3$[Co(CN)$_6$]$_2$ 与 GO 复合后在氮气下 600 ℃ 煅烧 2 h 获得了 Co$_3$ZnC/RGO 复合纳米结构。该材料具有优异的 HER 性能[76]。Su 等利用 Zn-Fe-ZIF 纳米球制备了铁掺杂的碳纳米管。他们首先用溶剂热反应制备了 3 μm 左右的 Zn-Fe-ZIF 微米球，该结构与 Zn-ZIF 类似，而 Fe 均匀地分布在颗粒中。如图 1.4 所示，多壁碳纳米管通过在氮气下 900 ℃ 直接加热 Zn-Fe-ZIF 获得。纳米管的直径为 50～800 nm，其形貌显示为竹节状，管两端未封口。通过对碳纳米管的生长过程分析，作者认为纳米管的生长过程类似于化学气相沉积（CVD）中的催化生长机理[77]。

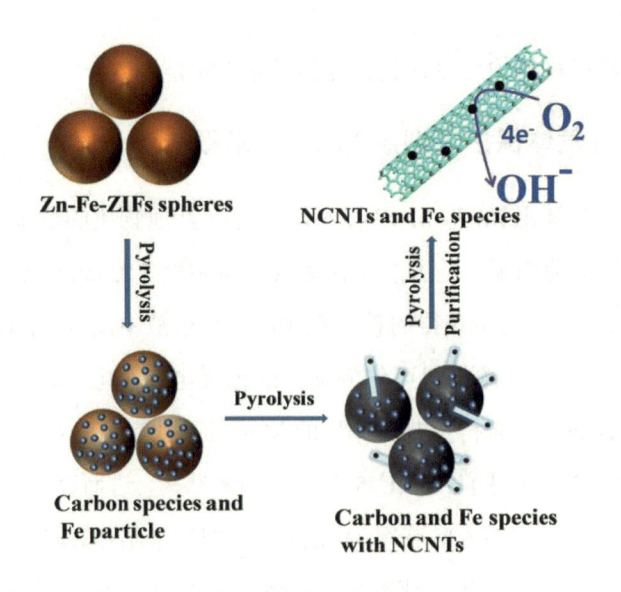

图 1.4　Zn-ZIF 上生长氮掺杂碳纳米管（NCNT）的过程[77]

Wu 等在 2013 年综述了利用含 N、C 有机物和 Fe、Co 等过渡金属在惰性气氛下催化生长氮掺杂碳材料，及其 M-N-C 催化生长机理。随后其课题组利用吸附 Fe 离子的 MOF 制备了不同直径氮掺杂石墨烯纳米管和石墨烯。首先他们制备了多孔结构的 Co-MOF，而后将二氰胺和醋酸铁吸附到 Co-MOF 中。混合物在氮气中加热，通过控制温度可分别得到洋葱状碳/铁纳米结构（800 ℃）、氮掺杂碳纳米管（900 ℃）和氮掺杂石墨烯-碳纳米管（1000 ℃）等碳材料。该材料具有优异的氧还原活性，可作为碳阴极材料用于 Li-O 电池。随后，他们利用 MIL-100（Fe）制备了竹节状的氮掺杂石墨烯纳米管（见图 1.5），并通过 Pt 掺杂获得了具有卓越氧还原性能的电极材料[78-80]。

2016 年，Zou 等通过两步煅烧 Ni-MOF 巧妙地获得了 NiO/Ni/GN 中空纳米结构。他们首先制备了均匀的 Ni-MOF 微球，接着在氩气保护下 450 ℃加热 30 min 获得 Ni-C 核壳结构，之后在空气中 200 ℃加热 30 min，使外壳的部分碳氧化，Ni 转化为 NiO，形成 NiO/Ni/GN 中空纳米结构。该材料可作为钠离子电池的负极材料（SIBs）[81]。

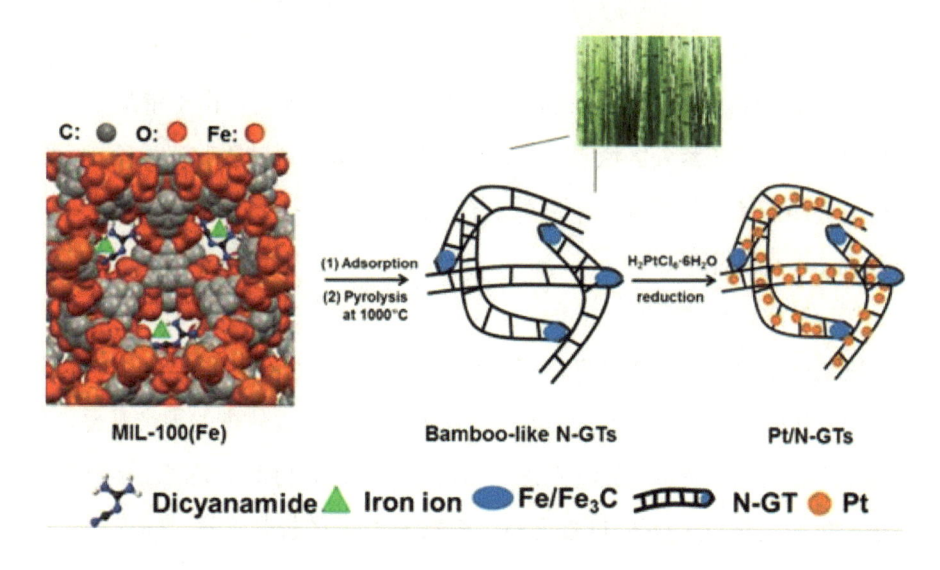

图 1.5 竹节状的氮掺杂石墨烯纳米管的生长过程[80]

1.4 中空/多孔纳米结构的应用

本书主要就中空/多孔纳米材料的光催化和电催化析氢性能（HER）进行研究，因此这里主要介绍光催化和 HER 相关背景及研究进展。

1.4.1 光催化

1.4.1.1 光催化原理

半导体光催化剂的能带在价带（Valence Band，VB）和导带（Conduction Band，CB）之间存在一个禁带（Forbidden Band，Band Gap）。由于半导体的光吸收阈值与带隙具有式 λ（nm）= $1\,240/E_g$（eV）的关系，当能量高于半导体吸收阈值的光照射半导体时，半导体的价带电子发生跃迁，即从价带跃迁到导带，从而产生光生电子（e^-）和空穴（h^+）。吸附在光催化剂颗粒表面的溶解氧俘获电子形成超氧负离子，而空穴将吸附在催化剂表面的氢氧根离子和水氧化成羟基自由基。而超氧负离子和羟基自由基具有很强的氧化性能，能将绝大多数的有机物氧化至最终产物 CO_2 和 H_2O[82]。

1.4.1.2 纳米复合物光催化剂

常见的单组分光催化剂，如 TiO_2 其光催化效率较低，由于带隙较宽，其吸收波长主要在紫外区域[83]。多组分复合可扩大材料的吸收光谱范围，同时提高界面载流子的转移速率常数，有效地抑制光生电子-空穴的复合，从而改善光催化性能[84]。

1. 离子掺杂

在催化剂晶格中引入掺杂离子可引入中间能级，降低半导体催化剂的带隙。TiO_2 晶格中掺杂金属元素后其 d 轨道和钛离子的 d 轨道的导带重叠，使其导带下移，禁带变窄，吸收光谱红移，对可见光的利用率提高。掺杂的阴离子产生的能级与 TiO_2 的价带重叠，相当于 TiO_2 的价带上移，同样使禁带变窄。掺杂离子还可成为电子和空穴的浅势捕获阱，抑制光生电子和空穴复合，延长受激载流子的寿命，大幅减少电子和空穴对的表面复合，增强其光催化活性。如果金属离子进入 TiO_2 的晶格取代了 Ti 的位置，或非金属原子取代氧原子的位置，产生了局部晶格畸变或形成了新的氧空位。Ti^{3+} 还原中心和氧空位可以充当反应的活性位置，氧空位数量的增加也会使光吸收红移至可见光区[85]。

2. 贵金属掺杂

贵金属的掺杂会改变半导体中电子的分布。如 TiO_2 和贵金属在表面接触时，载流子重新分布，电子从费米能级较高的半导体转移到费米能级较低的贵金属，直到它们的费米能级相匹配。在二者接触后形成的空间电荷层中，贵金属表面获得过量的负电荷，半导体表面显示出过量的正电荷，导致能带向上弯曲形成肖特基势垒，肖特基势垒能有效地捕获光生电子，阻止了光生电子与空穴的复合[86-87]。

半导体光催化剂中，贵金属的掺杂量存在最佳值。当贵金属沉积在 TiO_2 表面时，少量的贵金属掺杂产生的肖特基势垒不能阻止光生电子空穴对的复合。贵金属掺杂过量，会在 TiO_2 表面堆积，占据催化活性位点和比表面积，降低其光催化活性。当贵金属进入 TiO_2 晶格中置换 Ti^{4+} 时,贵金属掺杂量低,不能产生足够的杂质能级来捕获光生电子;贵金属掺杂量过大,杂质能级会成为光生电子空穴对的复合中心,降低催化剂的光催化活性。崔鹏等用不同 Ag 含量的 Ag/TiO_2 催化剂降解低浓度甲基橙溶液,$AgNO_3$ 溶液初始用量从 0.025%增加到 0.4%时光催化降解能力随 $AgNO_3$ 用量增加而逐渐增加。在 $AgNO_3$ 用量为 0.4%时,催化剂的效率最高。而当 $AgNO_3$ 用量达 0.6%时,催化剂的光催化性能反而降低。Pt/TiO_2、Au/TiO_2、Pd/TiO_2 催化剂,有类似的现象[88]。

3. 半导体复合

半导体复合实质上是一种半导体对另一种半导体的表面修饰。由于两种复合半导体的能带位置的不同,可能会形成如图 1.6 所示的两种类型的能带结构,分为 I 型能带结构和 II 型能带结构[89]。

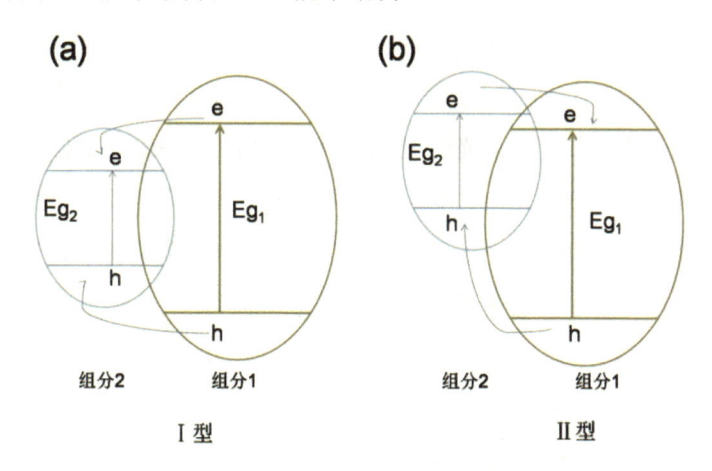

图 1.6 复合半导体(a)I 型、(b)II 型能带结构示意图[89]

I 型能带结构中(见图 1.6(a)),组分 1 的导带比组分 2 的导带略高些,而组分 1 的价带低于组分 2 的价带,使组分 1 的禁带宽度 Eg_1 大于组分 2 的禁带宽度 Eg_2。当入射光子能量大于或等于 Eg_2,但小于 Eg_1 时,只有组分 2 的电子被激发,而由于组分 1 和组分 2 的这种特殊的能带位置关系,光生电子-空穴对只能留在组分 2 能带中,无法向组分 1 转移。当光子能量大

于 E_{g1} 时，组分 1 和组分 2 同时被激发，由于静电势的作用，组分 1 导带的带负电荷的电子向组分 2 的导带转移，而组分 1 价带带正电荷的空穴向组分 2 的价带转移，组分 1 的电子和空穴均向组分 2 注入，在组分 2 内复合发光，对于发射与相关波长的光子是有利的，而从光催化机制的角度来讲，这一型的复合半导体并不适合作光催化剂。

II 型能带结构中（见图 1.6（b）），组分 1 导带低于组分 2 的导带，组分 1 的价带同样低于组分 2 的价带，二者的禁带宽度 E_{g1}、E_{g2} 的大小关系不确定。假如 E_{g1} 大于 E_{g2}，当入射的光子能量大于或等于 E_{g2} 但小于 E_{g1} 时，组分 2 的电子吸收能量跃迁至导带，并在价带上形成空穴，此时组分 1 不能被激发。在静电作用下，组分 2 的电子向导带势较低的组分 1 转移，充分地将光生电子和空穴分离开来。这一过程有利于电子-空穴在降解过程中发挥氧化还原作用。也就是说，当光子能量不小于 E_{g2} 时就可发生光催化反应。当入射的光子能量大于 E_{g1} 时，组分 1 和组分 2 均被激发，由于静电势作用，组分 2 的光生电子向导带势较低的组分 1 转移，组分 1 的光生空穴向价带势较高的组分 2 转移，光生电子与空穴分离。如果组分 2 相应的吸收边在可见光范围，组分 1 相应的吸收边在紫外光范围，那么组分 2 在这里担当了可见光敏化剂的角色，拓展了组分 1 吸收光子波长的范围，充分提高了光子利用率，相比传统的敏化活性剂，无机的半导体组分 2 往往展示出更优秀的光稳定性和更长的电子寿命。$CdS-TiO_2$、ZnO/CdS 均具有这样的结构[90-92]。

4. 局域表面等离子共振

当入射光照射在金属纳米颗粒上时，振荡电场会使传导电子一起振荡。金属表面存在的自由振荡的电子与光子相互作用，产生沿着金属表面传播的电子疏密波——表面等离子体。当入射光子与金属内的等离子体振荡的频率一致时就会产生共振，对入射光产生很强的吸收，发生局域表面等离子体共振（Localized Surface Plasmon Resonance，LSPR）效应。这一现象导致金属纳米颗粒呈黑色，能吸收可见光。由于共振的频率与金属粒子的微观结构特性密切相关，可通过调节金属纳米粒子的尺寸和形貌来调节 LSPR 效应。贵金属与半导体复合时，可以通过 LSPR 效应增强对入射光的吸收，而且可有效抑制光生电子-空穴的复合，大幅提高光催化材料的光转化效率[93]。

Huang 等提出了基于 LSPR 效应的 Ag@AgCl 纳米复合物光催化机理（见图 1.7）[94]。AgCl 颗粒的表面暴露出 Cl^-，其表面显负电性。表面的 Ag 纳米颗粒的电荷会在 AgCl 表面负电荷的作用下重新分布。Ag 内部的正电荷会分布在靠近 AgCl 的区域，而负电荷分布在远离 AgCl 表面的区域。在可见光的作用下，Ag 纳米颗粒的局域表面等离子体共振效应使得该类材料对可见光区有很强的吸收。产生的光电子迅速传递到 Ag 纳米颗粒的表面，而空穴则迁移到 AgCl 表面，并与 Cl^- 离子作用生成 Cl 原子。因 Cl 原子具有极强的活性，故能将有机物氧化。与此同时，Cl 原子又被还原为 Cl^- 离子，维持了体系的平衡。而光生电子并未传递到 AgCl，而是被体系中的 O_2 捕获，生成 O_2^-。这一过程确保了 Ag@AgCl 体系的稳定，从而改变了认为卤化银因稳定性差难以用于光催化反应的传统观念。

图 1.7　Ag@AgCl 光氧化甲基橙示意图[94]

1.4.1.3　中空/多孔纳米结构光催化剂

中空/多孔纳米结构光催化剂可以提供更多的催化活性位点，因此光催化效率高于其实心结构。Wu 等在乙二醇/水体系中制备了粒径小于 100 nm 的 SnO_2 中空纳米球。SnO_2 中空纳米球的比表面积达到了 71.232 $m^2·g^{-1}$，这使得其对罗丹明 B（RhB）的降解速率（一级动力学常数 $k = 0.012\ 1\ \text{min}^{-1}$）大于实心颗粒（$k = 0.010\ 8\ \text{min}^{-1}$）[95]。Kim 等用电纺法制备了中空结构的 $CoFe_2O_4$-聚苯胺纳米纤维，对甲基橙（MO）的降解速率常数为 $k = 0.016\ 1\ \text{min}^{-1}$ [96]。Chen 等用微波辅助的离子液体制备了 Ti^{3+} 掺杂的 TiO_2 中空纳米晶。该纳米晶对亚甲基蓝（MB）的降解速率常数达到了 $k = 0.037\ \text{min}^{-1}$ [97]。

Wang 等制备了一种具有可见光催化性能的 Bi_3NbO_7 纳米格子。Bi_3NbO_7 纳米格子可以在尿素作用下形成多孔结构，并且其对 RhB 和水杨酸的光催化效果随着尿素用量的增加而增强[98]。Chen 等在多孔结构的 Ga-In 双金属氧化物纳米光催化剂（见图 1.8）的超薄孔壁上发现了独特的电子结构，这一材料的导带和价带分布在两个相对的面上，并由一个小的静电电位差分离。这不仅缩短了光生电子从其产生的活性位点转移到发生催化反应位点之间

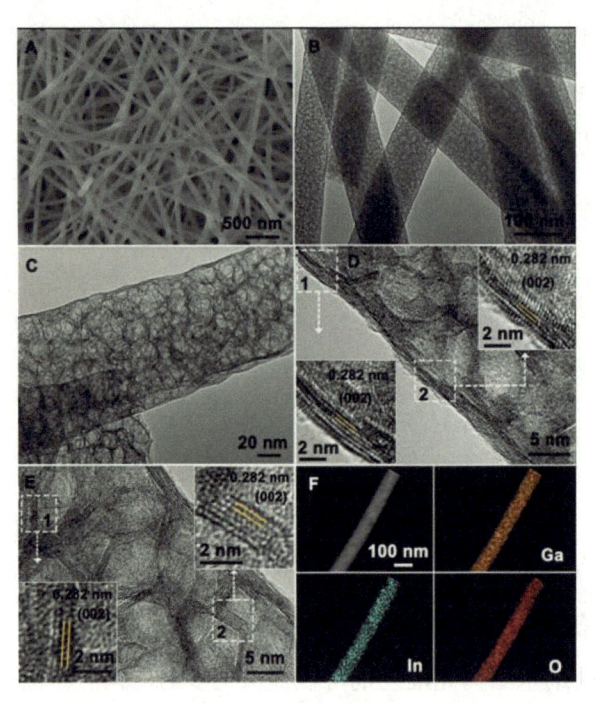

图 1.8 多孔的 $Ga_{1.7}In_{0.3}O_3$ 纳米结构的（a）SEM、（b，c）TEM、（d，e）HRTEM、（f）扫描透射电子显微镜（STEM）和元素 Mapping 图[99]

的距离,并且有利于材料中电荷的分离。这一多孔结构还导致了大量的表面反应/催化位点暴露。这一理想的电子和表面结构使得该材料表现出优异的光催化和电化学析氢性能[99]。这一结果为中空/多孔纳米结构在光、电领域的多功能应用提供了新的途径。

1.4.2 电化学析氢反应(HER)

水电解制氢是目前对环境污染最小的制氢技术[100]。然而,这一技术的成本过高导致其实际应用是有限的。尽管水电解技术历史悠久,但为了大幅降低其成本,持续的技术改进和材料创新仍然是非常必要的。随着技术的进步,最终水电解制氢技术将进入实际应用的阶段。Pt 基材料是 HER 催化剂的高效催化剂,但其价格昂贵。为了降低成本,高效 HER 催化剂最好由地球上丰富的元素制备[101-103]。在这样的背景下,研究人员一直在探索无贵金属 HER 催化剂。在过去的几年里,人们利用纳米技术制备了大量非贵金属催化剂,包括金属硫化物、金属氧化物、金属碳化物、金属氮化物、金属磷化物和杂原子掺杂的碳纳米材料。这里我们就 HER 电催化剂的工作原理、制备方法、催化活性进行介绍。

1.4.2.1 HER 的工作原理

电解水的电解槽由三个组成部分组成[104]:电解质(即 H_2O 溶液)、阴极和阳极。析氢催化剂和析氧催化剂分别覆盖在阴极和阳极。当外部电压施加到电极,水分子被分解成氢和氧。水分解反应可分为两个半反应:水氧化反应(析氧反应)和水还原反应(析氢反应)。在不同的介质中,水的分裂反应的表示方式不同。

总反应:$H_2O \rightarrow H_2 + 1/2O_2$

酸性介质中

阴极:$2H^+ + 2e^- \rightarrow H_2$

阳极:$H_2O \rightarrow 2H^+ + 1/2O_2 + 2e^-$

中性或碱性介质中

阴极：$2H_2O + 2e^- \rightarrow H_2 + 2OH^-$

阳极：$2OH^- \rightarrow H_2O + 1/2O_2 + 2e^-$

无论在何种介质中，在 25 ℃ 和 1 个大气压，水分解的热力学电压均为 1.23 V。另外，水分解的热力学电压是温度依赖性的，它随电解温度的增加而减小。事实上，为了实现电化学分解水我们必须将电压高于热力学势的值（1.23 V）。多余的电压（也被称为过电位）主要是用来克服激活两电极的内在势垒η_a和η_c，以及一些其他的电阻η_{other}，如溶液的电阻和接触电阻。因此，实际工作电压（E_{op}）为：

$$E_{op} = 1.23 \text{ V} + \eta_a + \eta_c + \eta_{other}$$

从这个方程可知，采用合适的方法降低过电位是解决电解水制氢效率问题的核心。η_{other}可以通过电解槽的优化设计减少，而η_a和η_c的最小化必须分别采用高活性的析氧和析氢催化剂。除了电极材料，电极的有效面积是决定反应的过电位的另一个重要因素。电极有效面积的提高可以通过电极的制备方法的优化来实现（例如纳米结构）。

在酸性介质中，HER 一般都是经历三个可能的反应步骤，而在碱性介质中的作用机制仍不清楚。在酸性介质中，第一个是所谓的 Volmer 步骤：$H^+ + e^- \rightarrow H_{ads}$。电子与质子在电极表面反应生成吸附的氢原子（$H_{ads}$）。之后，析氢反应可以继续进行由塔菲尔步骤（$2H_{ads} \rightarrow H_2$）或 Heyrovsky 步骤（$H_{ads} + H^+ + e^- \rightarrow H_2$）。不管 HER 通过何种途径进行，$H_{ads}$总会出现。因此，对氢的吸附自由能（$\Delta G_{H_{ads}}$）是被广泛接受的一种评价析氢材料性能的参数。例如，ΔG_H近似为零的 Pt，和类似 Pt 的固态析氢催化剂。如果$\Delta G_{H_{ads}}$过大且为正时，H_{ads}会与电极表面牢固结合，使 Volmer 步骤容易发生，但随后塔菲尔或 Heyrovsky 步骤变得困难。如果ΔG_H过小且为负时，H_{ads}具有与电极表面的弱相互作用，导致一个缓慢的 Volmer 步骤，并限制整体反应速率。因此，一个最佳的非 Pt 催化剂也应该提供适当的 HER 表面性质和有一个几乎为零的ΔG_{Hads}。

为了表征HER电催化剂的催化活性，需要测试并计算一些重要的参数。它们主要包括电极总活性、塔菲尔曲线、稳定性、电流效率以及循环次数。

1.4.2.2 非贵金属HER催化剂

这些元素，根据一般的物理和化学性质，大致分为三类：（1）贵金属铂（Pt）的HER催化剂；（2）过渡金属，用于构建的非贵金属催化剂，主要包括：铁（Fe）、钴（Co），镍（Ni）、铜（Cu）、钼（Mo）、钨（W）；（3）非金属材料，主要包括了硼（B）、碳（C）、氮（N）、磷（P）、硫（S）、硒（Se）。到目前为止，几乎所有的具有高效HER催化活性的非贵金属催化剂均由以上12种元素组成[105]。通过分析用于HER催化剂的金属的地壳丰度，可以得出以下结论：（1）铂的丰度约为$3.7 \times 10^{-6}\%$，明显比其他非贵金属的数量级小。这导致贵金属铂催化剂的成本最高。（2）6个非贵金属的地壳丰度为W = Mo < Co < Cu < Ni<<Fe，显然元素的地壳丰度越高，用于制备HER催化剂的成本越低。

12种元素中铁和镍的丰度最高，具有最低的价格，因此发展有效的铁或镍基催化剂是极有前景的。巧合的是，这一有挑战性的目标恰恰存在于生物体中——氢化酶。它广泛存在于自然界，催化氢质子和电子的交换，其活性位点都含有铁或镍离子[106]。因此选择地壳丰度高的铁、镍元素制备高效的纳米材料是提高HER技术效率并降低成本的有效途径。

1.4.2.3 中空/多孔纳米结构HER催化剂

Callejas等人制备了MnP型的FeP纳米粒子（约11 nm）。他们首先在油胺和十八碳烯的混合物中190 ℃分解$Fe(CO)_5$，随后在三正辛基膦（TOP）中340 ℃反应1 h（见图1.9）[107]。Deng等将Fe和Co封装在多壁碳纳米管中，获得了高效的HER催化剂[108]。Fang等通过煅烧Ni-MOF获得了Ni@C纳米结构，镍纳米颗粒封装在石墨壳中[109]。Feng等利用$Ni_3[Co(CN)_6]_2 \cdot 12H_2O$为前驱体通过煅烧和掺杂制备了镍钴磷化物中空纳米粒子，该粒子可作为在碱性溶液中的 HER 高效催化剂[110]。

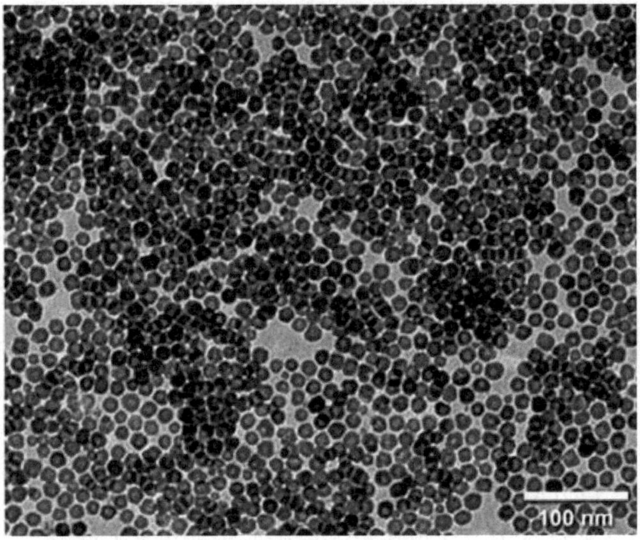

图 1.9　FeP 纳米颗粒的 TEM 图[107]

1.5　本书研究的内容

基于中空/多孔纳米结构在制备和应用方面的研究现状，本书设计并制备了多种二元和三元中空/多孔功能纳米材料，如 Ag/AgCl、$Zn_3[Fe(CN)_6]_2·xH_2O$、$Ag/ZnO/ZnFe_2O_4$、Fe/Fe_3C@N-doped C、NiFe 氮掺杂石墨烯微管、CoFe@C 等。通过对上述中空/多孔纳米结构制备方法和形成机理进行系统分析，探索这类材料的普适性制备方法。进而就这些材料在光催化、吸附、HER 领域的性能及其潜在应用作了进一步研究。本研究可为中空/多孔纳米结构的可设计制备和应用提供实验依据。

第 2 章　中空 Ag/AgCl 块状纳米结构的制备及其光催化性能研究

2.1 引言

由于贵金属纳米颗粒局域表面等离子共振（LSPR）效应及卤化银的光敏特性，Ag/AgCl 可被用于新型可见光催化剂[94]。研究人员已通过各种途径制备了不同形貌的 Ag/AgCl 纳米颗粒[111-112]，如类球形[113]、块状和准块状[114-119]、纳米线[120-121]，以及一些高活性晶面暴露的 AgCl 纳米晶[122-123]。但是由于银离子与氯离子反应速率过快，导致制备特定形貌 AgCl 纳米颗粒的过程比较难以控制[124]。

具有中空或多孔结构的材料与实心材料相比具有更大的比表面积，这一点使它们在催化剂、吸附和药物传输等领域得到应用[125-127]。尽管目前制备中空纳米结构的途径很多，但是 AgCl 中空颗粒的制备研究较少[128-134]。最近 Tang 等利用 NaCl 晶体作为模板制备了 Ag@AgCl 中空方块，并研究了其卓越的光催化性能[111]。本章报道一种简单的一锅法制备中空 AgCl 纳米结构，并尝试用溶解-沉淀与离子扩散相结合的机理来解释中空纳米结构的制备过程[135-136]，进而通过光辐照 AgCl 纳米结构获得了 Ag/AgCl 光催化剂，该催化剂对有机污染物表现出卓越的光催化性能。

2.2 材料和方法

2.2.1 材料

本实验用到的所有化学试剂均为分析纯，购自国药集团。实验中用到的水均为去离子水。

2.2.2 AgCl 中空方块的制备

制备过程中用 $AgNO_3$ 和 CCl_4 作为前驱体。30 mg $AgNO_3$ 和 200 mg PVP 溶解在 40 mL 无水乙醇中，然后加入 20 mL 含有 8 mg NaOH 的乙醇溶液，并搅拌至溶液呈暗红色。取暗红色溶液 12 mL 与 10 mL 的 CCl_4 溶液混合均匀后转移到 25 mL 聚四氟乙烯内胆的反应釜中，密封，并在 93℃ 下反应 4 h。反应釜冷却到室温后，离心分离 AgCl 沉淀，分别用乙醇和去离子水洗 5 次。产品分散到少量乙醇中避光保存，用前在 45℃ 真空烘箱中避光干燥。为了获得 Ag/AgCl 催化剂，我们将 AgCl 样品分散到 10 mL 去离子水中，用 500 W 钨灯分别辐照 10 min（记为 Ag/AgCl-1）、20 min（记为 Ag/AgCl-2）和 30 min（记为 Ag/AgCl-3）。

2.2.3 仪器

物相结构由 X 射线衍射（XRD，Rigaku D/MAX 2400，Cu-K_α λ = 1.540 6 Å（1 Å=0.1 nm），扫描速率 2 °/min）表征。样品的形貌、粒径和微观结构用 FESEM（Hitachi S-4800，加速电压 20 kV）和 TEM（JEM-2100，加速电压 200 kV）表征。紫外可见光谱均由 UV-2450 紫外-可见分光光度计检测。将 0.1 g Ag/AgCl 溶解在 25 mL 的 1 $mol·L^{-1}$ 的氨水中，离心分离后，用离子色谱（ICS-1100，Dionex Ionpac™ AG11-HC，流速：1.0 $mL·min^{-1}$，温度：30 ℃，压力：1 604 psi（1 MPa=145 psi），SRS 电流：8 mA）测试溶液中 Ag^+ 浓度。

2.2.4 光催化性能

样品的光催化性能用在模拟太阳光下降解甲基橙（MO）的方法来测试。500 W 的钨灯（或 250 W 氙灯）置于石英套管反应器 10 cm 处。套管夹层中通冷凝水，用于降低钨灯辐射产生的热量。50 mg 光催化剂加入 100 mL 浓度为 10 $mg·L^{-1}$ 的 MO 溶液中。在辐照前样品在避光条件下搅拌 30 min，使催化剂与染料达到吸附平衡。吸附平衡后开始光催化反应。到达设定的反应时间后抽取 4 mL 悬浊液，离心分离除去催化剂。上层清液中 MO 的浓度用紫外-可见分光光度计在 505 nm 处测量。总有机碳（TOC）用 TOC 分析仪（Analytik Jena AG multi N/C 2100）测定。

2.3 结果与讨论

2.3.1 结构与形貌

图 2.1（a）给出了制备 AgCl 样品的 XRD 谱图。图中所有的衍射峰与立方 AgCl 标准图谱（JCPDS No：31-1238）一致，表明样品为纯净的 AgCl。图 2.1（b）为 AgCl 中空结构的 FESEM 图谱，图中可以看出大量均匀的块状颗粒，每个颗粒的中间有一个大的空腔结构。从高倍 FESEM 图（2.1（c）和 2.1（d））可以看出颗粒的尺寸为 600～900 nm。AgCl 方块的中空结构可以用 TEM 表征。从图 2.1（e）和 2.1（f）可以看出颗粒的壁厚小于 100 nm，空腔约 400～600 nm。由于在高能电子辐射下 AgCl 会分解，在颗粒的表面可以看到一些银的纳米颗粒。

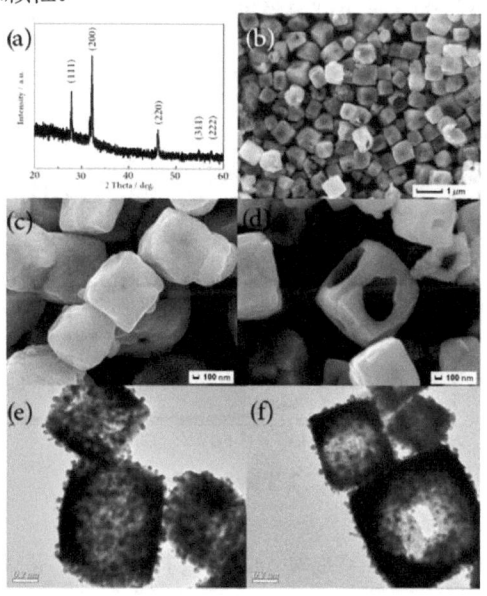

图 2.1 AgCl 中空方块的（a）XRD 图谱、（b，c，d）FESEM 和（e，f）TEM 图

2.3.2 AgCl 中空方块的形成过程

通过不同反应时间下得到的产品的形貌可以研究 AgCl 中空方块的生长过程。首先 AgNO$_3$ 与 NaOH 反应转化成 Ag$_2$O，并被乙醇还原成 Ag^0 晶种，溶液变成暗红色（见图 2.2（a））。对 Ag^0 晶种进行 XRD、SEM 和 TEM 表征。如图 2.3（a）所示，银晶种的 XRD 图中所有的衍射峰都与立方相的银的标准图谱（JCPDS No：65-2871）一致。银晶种的 TEM 和 HRTEM 图如图 2.4 所示，晶种呈类球状，晶格常数为 2.36 Å，这与面心立方银的{111}晶面相对应。

图 2.2 （a）Ag^0 晶种和（b）AgCl 的数码照片

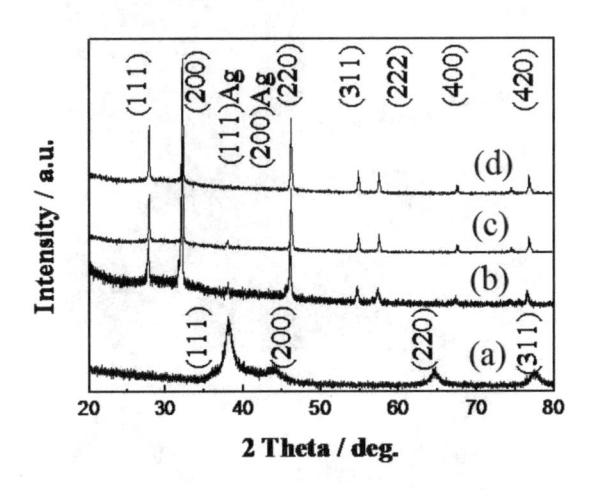

图 2.3 （a）Ag^0 晶种，（b）粉红色的 AgCl@Ag 样品，（c）花状和（d）准立方块状 AgCl 的 XRD 图谱

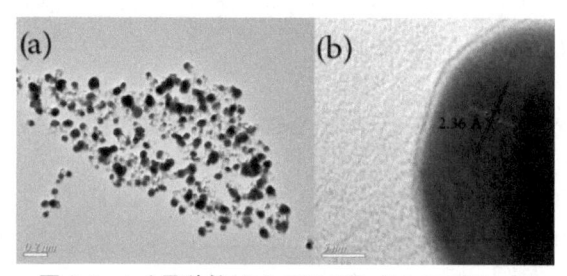

图 2.4 Ag⁰ 晶种的（a）TEM 和（b）HRTEM 图

银晶种与 CCl_4 反应 1 h 后，暗红色的溶液变为粉红色。样品的 FESEM 图如图 2.5（a）所示，图中可以看出银的晶核被一壳层所包覆。粉红色样品的 XRD 图（见图 2.3（b））中有两个弱衍射峰 $2\theta = 38.1°$（111）和 44.3°（200），可归属为金属银，其余的峰归属于 AgCl。这一结果说明银晶核上已经长出了 AgCl。随着反应时间增加，产物形貌变为花状的形貌（2 h，图 2.5（b））和类块状形貌（3 h，图 2.5（c）），同时颗粒的粒径变得更加均匀。图 2.5（c）中颗粒表面留下的凹痕可能是颗粒生长留下的痕迹。由于 AgCl 在电子束下快速分解，不能进行 HRTEM 分析，所以只能通过颗粒形貌的变化来解释晶体的生长[137]。通过 4 h 的溶剂热反应可以获得均匀的 AgCl 中空结构。如果延长反应时间至 4.5 h，如图 2.5（d）所示，颗粒内部的空腔将增大，产生大量的框架结构。

图 2.5 不同水热反应时间下制备的样品的 FESEM 图：（a）1 h、（b）2 h、（c）3 h、（d）4.5 h

2.3.3 AgCl 中空方块的生长机理

根据上述实验结果，可以认为 AgCl 中空方块的生长过程可能分为三步（见图 2.6）：

1. 金属银晶种的形成

银晶种在 $AgNO_3$ 和 NaOH 的乙醇溶液中形成。这里乙醇作为还原剂，因为在碱性条件下醇类物质可以将银离子还原为金属银[138-139]。反应过程如式（2.1）、（2.2）：

$$2Ag^+ + 2OH^- \longrightarrow Ag_2O + H_2O \qquad (2.1)$$

$$C_2H_5OH + Ag_2O \xrightarrow{OH^-} CH_3CHO + 2Ag^0 + H_2O \qquad (2.2)$$

同时，Ag_2O 和 Ag^+ 可以被（2.2）中生成的乙醛还原为金属银 Ag^0。

$$CH_3CHO + 2Ag^+ + 2OH^- \longrightarrow CH_3COOH + 2Ag^0 + H_2O \qquad (2.3)$$

$$CH_3CHO + Ag_2O \xrightarrow{OH^-} CH_3COOH + 2Ag^0 \qquad (2.4)$$

2. AgCl 在银颗粒的 {111} 面生长[140]

$$(4-x)Ag + CCl_4 \rightarrow (4-x)AgCl + CCl_x \qquad (2.5)$$

当 $x = 0$ 时，方程式（2.5）的自由能 $\Delta_r G^0 = -374.0 \text{ kJ·mol}^{-1}$。该反应很容易进行，但反应速率较慢[141]。

在溶剂热反应中，Ag^0 的晶种容易聚集[142]。当加热时，为了降低表面能量，Ag^0 的晶种聚集形成更大的颗粒[143]，进而在银颗粒表面原位生长出一层 AgCl。随着反应的进行，银颗粒可以长成八面体形的颗粒[144]。

3. 溶解-沉淀机制

在后续的反应中，一些较小的颗粒会被较大的均匀颗粒"吃掉"，所以这一阶段的产物的形貌呈现规则的花状和块状，并且随着反应时间的延长颗粒更加规整。同时方块内核的银原子逐渐向外扩散，并且在表面与 CCl_4 反应，这里银原子被 CCl_4 氧化形成 AgCl 并沉积在颗粒表面，在 Ag^0 聚集体表面形成 AgCl 壳。目前普遍认为阳离子的扩散速率比阴离子要快[29-30]，本实验中银离子的扩散速度比氯离子要快,这导致立方块的中心出现孔洞,最终形成了中空结构。这一结果说明本实验中 AgCl 的中空结构是通过反应时间来控制的。

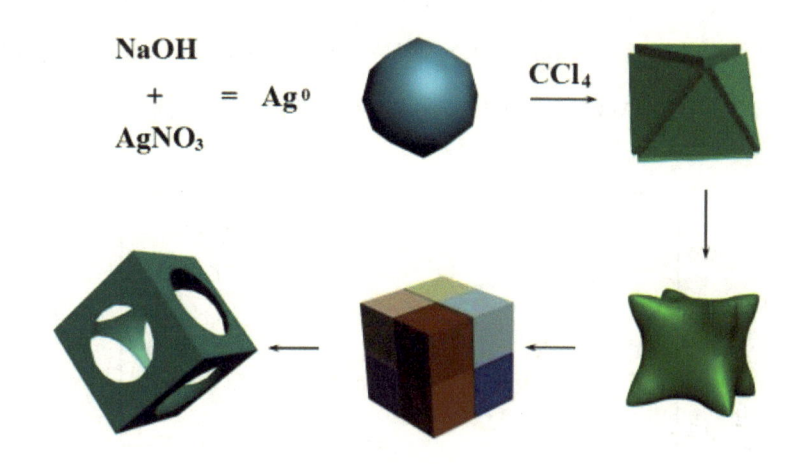

图 2.6 AgCl 中空块的形成过程

2.3.4 中空 Ag/AgCl 方块的光催化性能

用离子色谱法检测 Ag/AgCl 催化剂中的 Ag^+ 的含量。结果如表 2.1 所示,Ag:AgCl 可以通过调节辐照时间来控制,增加辐照时间中空 AgCl 方块表面的银颗粒逐渐增加。最佳的 Ag:AgCl 比例是 9.6:100(Ag/AgCl-2)。但是进一步增加银的比例,颗粒的粒径和 Ag/AgCl 的接触面会增加,带有负电荷的银颗粒表面会俘获光生空穴,这将降低电子的分离效率,导致光催化效率下降[145-146]。

表 2.1 Ag/AgCl 催化剂中 Ag 原子价态分析

样品	Ag:Cl	$Ag^0:Ag^I$
Ag/AgCl-1	51.2:48.8	4.9:100
Ag/AgCl-2	52.3:47.7	9.6:100
Ag/AgCl-3	53.1:46.9	13.2:100

目前，普遍认为光吸收强度在光催化反应中扮演重要角色，尤其是对于有机污染物的降解过程[147-148]。太阳光催化的效率依赖于催化剂对太阳光的利用率。AgCl 中空方块和 Ag/AgCl 复合方块的光吸收性质通过紫外-可见漫反射方法测定，结果如图 2.7 所示。AgCl 由于带隙较大主要吸收紫外光，在可见光区的吸收很弱。而 Ag/AgCl-2 样品在 400～650 nm 有一个宽的吸收峰，这个吸收峰是由于 AgCl 表面的银纳米颗粒的 LSPR 效应引起的。

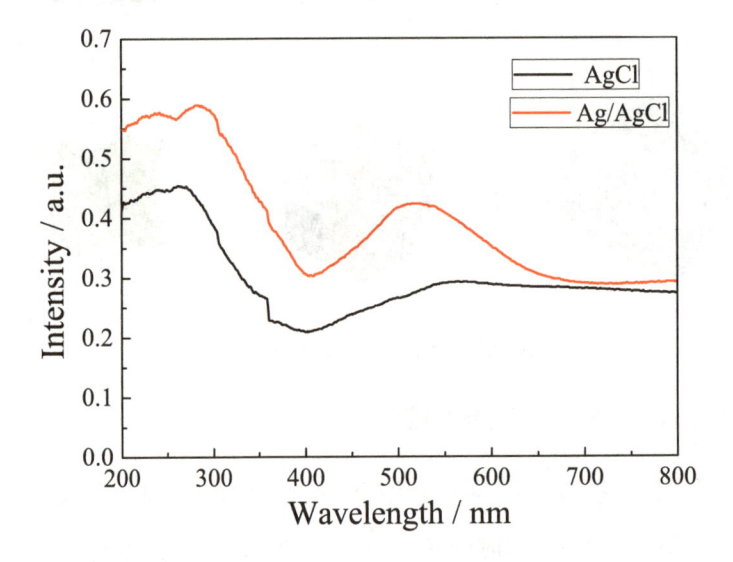

图 2.7 AgCl 和 Ag/AgCl-2 中空立方块的紫外-可见漫反射谱

用钨灯辐照降解 MO 作为模型来评价样品的光催化活性。用商业级的氮掺杂二氧化钛（T_{25}）作为对比。如图 2.8（a）所示，T_{25}、AgCl 和 Ag/AgCl 的光催化性能逐渐增强。Ag/AgCl 中空方块的光催化性能比 AgCl 中空方块强。辐照 30 min 后，AgCl 仅可以降解 50%的 MO，而 Ag/AgCl-2 的降解率可达到 97%。为了评价催化剂的稳定性和重复性，我们测试了 Ag/AgCl-2 在使用回收后的光催化效率。如图 2.8（b）所示，5 个循环后，催化剂的活性略有下降，说明催化剂的稳定性较好。

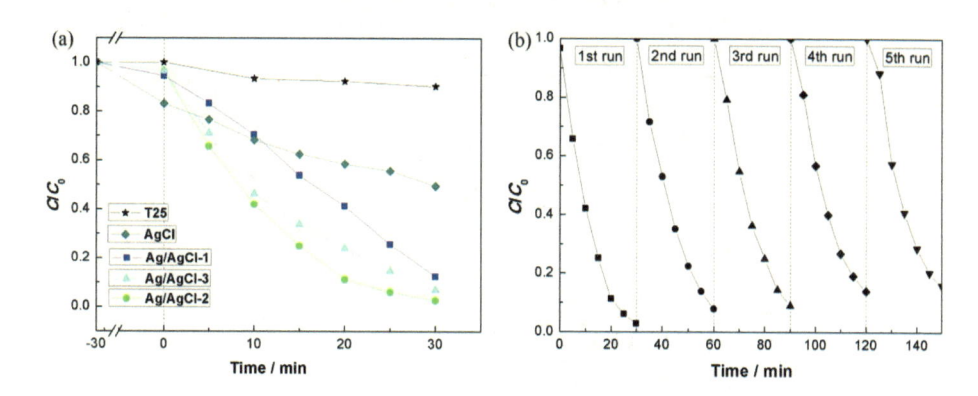

图 2.8 （a）T25、AgCl 和 Ag/AgCl 样品对 MO 的光催化降解曲线，
（b）Ag/AgCl-2 对 MO 降解的循环实验

TOC 分析可以测试催化剂对 MO 降解的矿化程度[149]。在钨灯丝辐射 30 min 后，AgCl、Ag/AgCl-1、Ag/AgCl-2 和 Ag/AgCl-3 的 TOC 去除率分别为 64%、74%、79% 和 77%。这一结果说明随着 MO 的分解，部分染料矿化为 H_2O 和 CO_2。这一结果进一步证明了 Ag/AgCl-2 的光催化活性最好。

光催化降解染料的反应可看作准一级反应（见图 2.9），用 ln（c/c_0）-t 作图，经拟合后获得一条直线。其反应速率常数 k（min^{-1}）为直线的斜率。对于中空结构的 Ag/AgCl-2 其速率常数 k 为 0.117 7 min^{-1}，这一结果比之前报道的实心 Ag/AgCl 复合物高[146]。为了比较 Ag/AgCl 和 Tang 等报道的 Ag@AgCl 中空复合结构的光催化性能，用 250 W 氙灯作为光源，考察了 Ag/AgCl-2 的光催化性能。实验结果如图 2.10 所示，经过 10 min 照射后，MO 溶液的降解率达到了 97%。催化剂在氙灯下的催化性能明显比在钨灯下高，其光催化活性与 Tang 等报道的结果相当（6 min，300 W 氙灯）[111]。这一结果进一步说明中空结构可以增强 Ag/AgCl 的光催化性能。这是由于中空结构可以为催化反应提供更多的活性位点[150-154]。

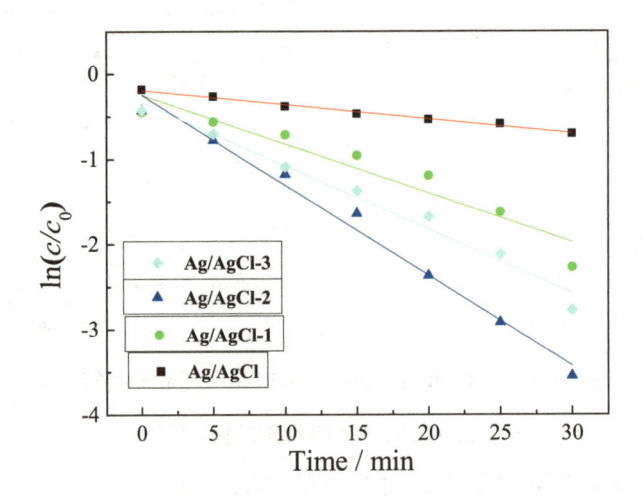

图 2.9 Ag/AgCl 和 AgCl 准一级动力学曲线 ln（c/c_0）-t

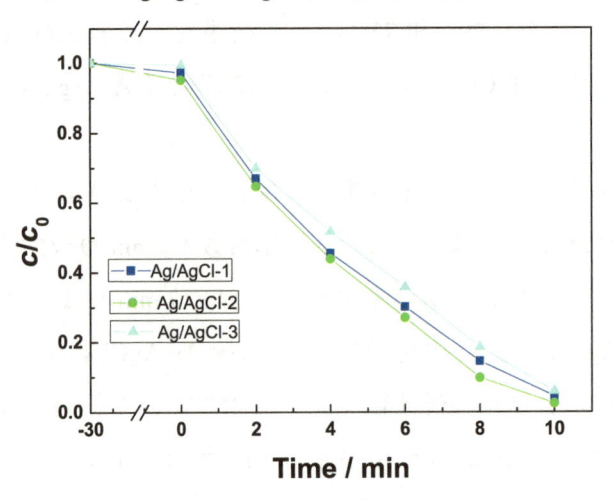

图 2.10 在氙灯辐照下 Ag/AgCl 中空块对 MO 的降解曲线

通常认为光催化过程中空穴（h^+）、羟基自由基（·OH）和超氧自由基（O_2^-）对于有机污染物的降解起到重要作用[155]。为了确定主要的催化活性物质并了解光催化机理，在光催化反应体系中添加了少量的三乙醇胺（TEOA，h^+ 俘获剂）、叔丁醇（t-BuOH，·OH 俘获剂）和对苯醌（BZQ，O_2^- 俘获剂)[156-157]。如图 2.11 所示，Ag/AgCl-2 在添加了 t-BuOH、

BZQ 和 TEOA 的体系中反应速率常数 k 分别为 0.075 6 min^{-1}、0.006 2 min^{-1} 和 0.003 5 min^{-1}。这说明在光降解 MO 的过程中添加 t-BuOH 没有明显降低催化剂的活性，而对苯醌和 TEOA 的加入明显降低了催化剂的活性。这说明 OH 不是该催化剂产生的活性成分，O_2^- 和 h^+ 对 MO 的光催化过程起到关键作用。

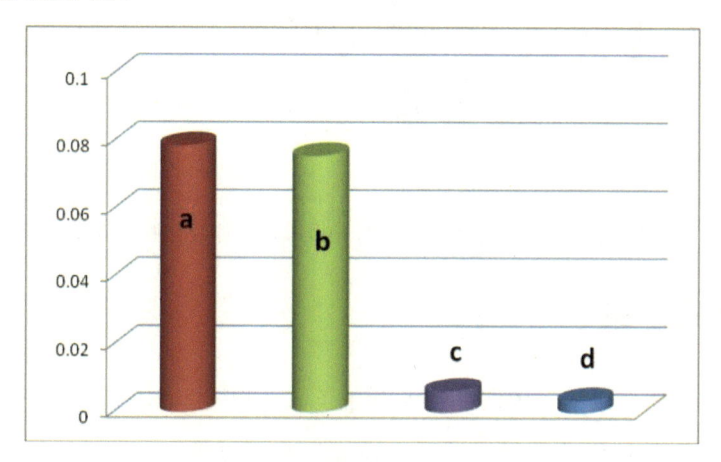

图 2.11 （a）Ag/AgCl-2 和反应体系中分别添加俘获剂（b）1 mM t-BuOH、（c）1 mM BZQ、（d）1 mM TEOA 后的光催化反应速率常数

由于纯 AgCl 的直接和间接带隙分别为 5.6 eV 和 3.25 eV，其只能吸收紫外光[158]。因此 Ag/AgCl-2 的可见光催化性能得益于其表面的 Ag 纳米粒子产生的表面等离子共振效应[94]。由于 Ag 纳米颗粒导电性和 AgCl 表面的负电性[159]，催化剂界面上产生的光生电子转移到 Ag 纳米颗粒。接着这些电子与注入的 LSPR 电子一起被溶液中的氧分子俘获，生成超氧自由基，用于氧化 MO 分子[160]。同时，空穴与 Cl^- 离子结合形成 Cl^0 原子。由于 Cl^0 原子的氧化性很强，Cl^0 原子将 MO 分子氧化，自身被还原为 Cl^- 离子[160]。催化机理图 2.12 描述这一过程。由于 Ag 纳米粒子是在 AgCl 表面原位生成的，在界面间的电子转移和光生电子-空穴的分离是光催化性能提高的另一主要因素[121]。

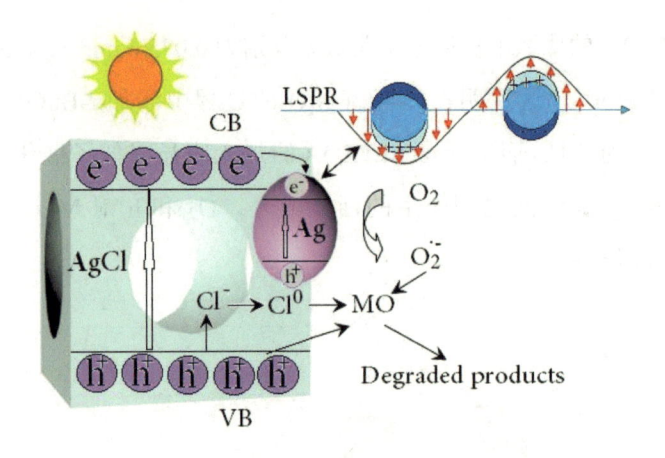

图 2.12　Ag/AgCl 中空方块对 MO 降解可能的光催化机理

2.4 本章小结

本章采用新颖的一锅溶剂热法制备了 AgCl 中空方块。研究了 AgCl 中空方块的形貌转化及其形成机理。通过光照可以将 AgCl 中空方块转化为 Ag/AgCl 复合光催化剂。由于 Ag 的 LSPR 效应，复合催化剂的光催化性能大幅提高。同时，提出了可能的光催化机理。该催化剂有望在光催化降解有机污染物、光伏电池等光电设备领域获得应用。

第 3 章　Ag/AgCl 纳米结构的形貌可控制备及其光催化性能研究

3.1 引言

近 20 年来制备具有特定尺寸和形貌的纳米结构一直备受关注[161]。通过控制纳米颗粒形貌和粒径可以使其获得广泛的应用[162]，这促使纳米颗粒的形貌可控制备得到了快速发展[163-167]。其中液相法是大量制备均匀形貌纳米颗粒的理想方法[168-169]。

由于 Ag/AgCl 纳米材料具有优异的光催化性能，其形貌可控制备受到了研究人员的极大关注[94, 113-114, 116, 119, 121]。目前 AgCl 纳米材料的制备方法主要有模板法、离子液体辅助法、表面活性剂法等[145, 159, 170-171]。通常结晶质量高的小粒径的半导体纳米晶体具有较高的光催化活性，但是目前形貌可控制备 100 nm 以下的高质量 AgCl 纳米晶依然是个挑战[98, 124]。

本章介绍一种一步法制备尺寸 100 nm 以下 AgCl 纳米晶的可控制备方法，并用沉淀-溶解机制解释了其生长机理。通过紫外线辐照在 AgCl 纳米晶的表面原位生长 Ag 纳米颗粒，形成 Ag/AgCl 复合光催化剂。该催化剂在太阳光下对罗丹明 B（RhB）具有较好的光催化降解活性，且具有较好的光催化稳定性。

3.2 材料和方法

3.2.1 材料

PVP（K-30，平均 M_w = 40 000）、AgNO$_3$、CCl$_4$、NaOH、二氯乙烷和乙醇均购于上海化学试剂公司。实验用水为二次蒸馏水。

3.2.2 制备方法

AgCl 纳米球：30 mg AgNO$_3$ 和 0.2 g 的 PVP 溶解在 8 mL 乙醇中。超声溶解后，将其转移到 20 mL 的反应釜的聚四氟乙烯内胆中，并加入 8 mL CCl$_4$ 搅拌均匀后，密封。反应釜在 120 °C 加热 1 h，冷却到室温。所得乳白色沉淀离心分离并用乙醇和水分别洗涤三次。之后沉淀在真空烘箱中 50 °C 烘干 24 h。其他条件不变，温度从 120 °C 增加到 160 °C 可以得到 AgCl 纳米块。在上述反应中 AgNO$_3$ 溶液中加入含 8 mg NaOH 的乙醇溶液 4 mL，并用二氯乙烷代替 CCl$_4$，反应温度 120 °C 保持 8 h，可得到 AgCl 四面体。

3.2.3 仪器

物相结构由 X 射线衍射（XRD，Rigaku D/MAX 2400，Cu-K_α λ = 1.5406 Å，扫描速率 2 °/min）表征。样品的形貌、粒径和微观结构用 FESEM（Hitachi S-4800，加速电压 20 kV）和 TEM（JEM-2100，加速电压 200 kV）表征。紫外可见光谱均由 UV-2450 紫外-可见分光光度计检测。

3.2.4 光催化性能

样品的光催化性能用在模拟太阳光下降解 RhB 溶液的方法来测试。250 W 的氙灯置于石英套管反应器 10 cm 处。套管夹层中通冷凝水，用于降低氙灯辐射产生的热量。50 mg 的光催化剂加入 100 mL 的 RhB 溶液中，RhB 溶液的浓度为 10 mg·L^{-1}。在辐照前样品在避光条件下搅拌 30 min，用于使催化剂与染料达到吸附平衡。吸附平衡后开始光催化反应。到达设定的反应时间后抽取 4 mL 悬浊液，离心分离去除催化剂。上清液中的 RhB 溶液的浓度用紫外-可见分光光度计在 554 nm 处测量。总有机碳（TOC）用 TOC 分析仪（Analytik Jena AG multi N/C 2100）测定。

3.3 结果与讨论

3.3.1 AgCl 纳米晶的形貌可控制备

AgCl 纳米方块、纳米球和四面体的 XRD 结果如图 3.1 所示。图 3.1 中所有样品的衍射峰均与立方相 AgCl 的标准卡片（JCPDS no. 31-1238）相一致，这表明三个样品均为纯净的 AgCl，不含其他杂质。值得注意的是，方块状样品的谱图（图 3.1（b））中衍射峰的相对强度与标准谱图不同，其衍射峰强度比 I_{200}/I_{111}（$ca.$ 12）要比标准谱图（2.0）中的大得多，这说明纳米方块沿着（200）晶面生长，每个立方块由 6 个（200）晶面构成。

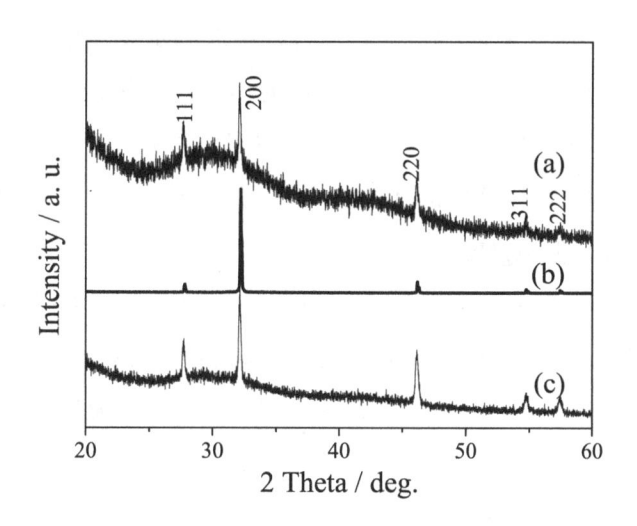

图 3.1 AgCl 样品的 XRD 谱图：（a）纳米球、（b）纳米块和（c）四面体

FESEM 显示（见图 3.2（a））120 °C 下获得样品的形貌为规则的球体，直径约 80 nm。而 160 °C 下获得的样品形貌为规则的去角立方块，边长约为 100 nm。AgCl 纳米颗粒在高压电子束下会快速分解，故未对其进行 TEM 和 HRTEM 分析。

图 3.2 AgCl 样品的 FESEM 图：（a）纳米球、（b）纳米块

3.3.2 反应温度和时间对形貌的影响

温度在 AgCl 纳米晶的形貌转化过程中起到重要作用。当反应温度低于 100 °C，反应 1 h 后只能得到黄色的样品，从其 SEM 图（见图 3.3（a））可看到该样品没有规则的形貌。从其 XRD 图谱（见图 3.3（b））可以证明该样品为 Ag 和 AgCl 的混合物。这一结果说明，在此温度下反应进行得很慢，但 AgNO$_3$ 可以分解产生银核，并与 CCl$_4$ 分解产生的 Cl 反应生成 AgCl。在 120 °C 下反应可以获得形貌较好的 AgCl 纳米球。如图 3.4（a）所示，反应温度升高到 130 °C 时制备的纳米颗粒开始出现棱角，随着温度升高，颗粒边缘变得越来越清晰。反应温度升高到 140 °C 后，从 FESEM 图中（见图 3.4（b））可以看到除了颗粒粒径变大，样品中还出现了一些块状的颗粒。这表明随着温度的升高，AgCl 颗粒的形貌逐渐从球形向块状转化。当温度升高到 160 °C 时，样品的形貌变为典型的块状（见图 3.2（b））。在 170 °C（见图 3.4（c））下获得的样品形貌与 160 °C 基本一致。但是温度超过 180 °C 后，样品的形貌变得不规则（见图 3.4（d）），且出现许多四棱柱形颗粒。这可能是因为反应温度超过 180 °C 后，PVP 容易聚合[172]，这导致样品的粒径变大，出现长的矩形颗粒。

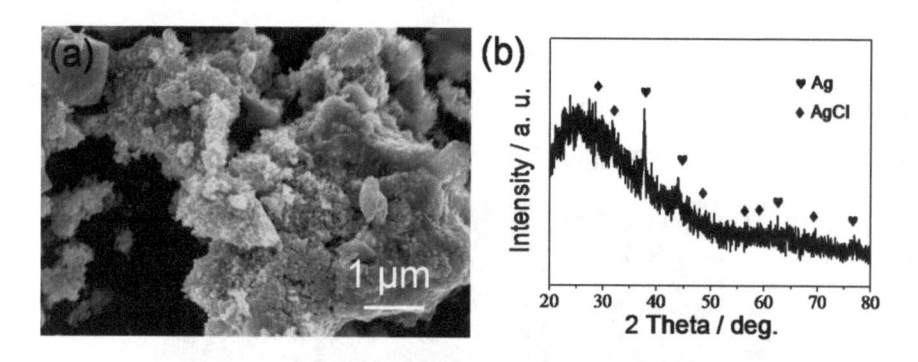

图 3.3 100 ℃下反应 1 h 制备样品的（a）SEM 和（b）XRD 图

图 3.4 不同温度下制备的 AgCl 样品的 FESEM 图片：（a）130 ℃、（b）140 ℃、
（c）170 ℃、（d）180 ℃

另一方面，除了反应温度，时间也是影响产物形貌的重要因素。在低温（120 ℃）和高温（160 ℃）下分别反应 2 h、4 h、6 h 和 8 h，，产物形貌会发生明显变化。图 3.5（a）是在 120 ℃下反应 2 h 获得样品的 FESEM 图片。如图所示，颗粒为规则的球形，其直径约 90 nm。随着反应时间继续增加到 4 h 和 6 h（见图 3.5（b）、（c）），产物的粒径逐渐变大，形貌也来越接近立方块状。当反应时间增加到 8 h，颗粒的形貌明显变为块状（见图

3.5（d）），其粒径增加到 100 nm。图 3.6 为 160 ℃下反应 2~8 h 后获得样品的 SEM 图片。如图所示，颗粒始终保持块状形貌，只是边缘变得越来越明显。其粒径也从 100 nm 增加到 500 nm。

图 3.5　160 ℃下不同时间制备的 AgCl 样品的 FESEM 图：（a）2 h、（b）4 h、（c）6 h、（d）8 h

图 3.6　160 ℃下不同时间制备的 AgCl 样品的 SEM 图：（a）2 h、（b）4 h、（c）6 h、（d）8 h

3.3.3 反应溶剂对形貌的影响

除了用 CCl₄ 作为氯源，还可以用二氯乙烷作为氯源制备 AgCl 纳米晶。其结果如图 3.7 所示。从图中可以看到 AgNO₃ 和二氯乙烷在 120 °C 下反应可以获得四面体型纳米颗粒，随着反应时间延长，颗粒的粒径变大，同时四面体所占的比例增加。反应 3 h 所得样品为立方块和四面体的混合物，直径约 200 nm。其中，立方块所占比例居大。随着反应时间延长，颗粒的粒径变大，同时四面体的比例逐渐增加，如图 3.7（b）～（e）。当反应时间延长到 8 h 时，颗粒形貌以四面体微块为主，同时粒径达到了 2 μm 左右（见图 3.7（f））。高倍的 FESEM 图片显示四面体形颗粒的表面比较粗糙，可以看到一些小颗粒。这些颗粒可能是 AgCl 分解产生的银。

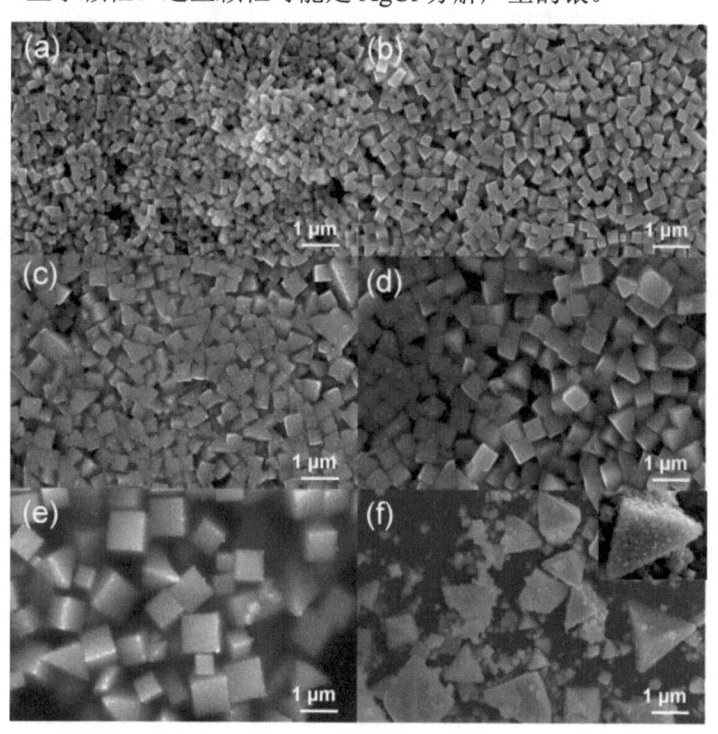

图 3.7　以二氯乙烷为碳源制备的 AgCl 样品的 SEM 图：（a）反应 3 h、（b）4 h、（c）5 h、（d）6 h、（e）7 h、（f）8 h

为了更好地理解 AgCl 纳米晶的生长过程,我们用沉淀-溶解机制解释纳米晶的形成过程。我们认为 AgCl 纳米晶的生长过程分为三个阶段,包括成核、晶种聚集和沉淀溶解过程。首先 CCl_4 缓慢地分解为 Cl^- 离子,然后 Cl^- 离子与 Ag^+ 离子反应生成 AgCl 晶核。这一过程中,CCl_4 的分解速率随温度升高而增加,而 Cl^- 离子浓度与 CCl_4 分解的速率相关。为了降低表面能,AgCl 晶核有聚集成球形颗粒的倾向[173-174]。因此 120 °C 下反应 1 h 只能获得较小的球形颗粒。随着反应时间的延长,反应体系发生奥斯特瓦尔德老化。依据奥斯特瓦尔德原理,小颗粒的溶解度比大颗粒大[142]。因此随着反应的进行,小颗粒逐渐消失,并沉积到大颗粒上,导致大颗粒逐渐长大,并最终形成块状颗粒[175-176]。当温度升高到 160 °C 时,反应速率加快,仅反应 1 h,就可以获得较大尺寸的立方块。

此外,以二氯乙烷替代 CCl_4,同时体系中加入 NaOH,可以获得四面体形纳米颗粒。由于四面体的对称性较低,面心立方结构的 AgCl 容易获得立方块或球形纳米颗粒,很难获得四面体形的纳米颗粒。获得四面体形的纳米颗粒只能通过控制晶体生长的反应动力学过程实现[177]。根据文献报道,四面体形纳米块可以通过减慢反应速率的方式获得[178]。这一过程可以通过控制前驱体浓度来实现[179]。如果以二氯乙烷代替 CCl_4,前驱体分解产生的 Cl^- 浓度降低。同时,加入 NaOH 可以将 $AgNO_3$ 转化为 Ag 从而降低 Ag^+ 的浓度。而 PVP 在该反应中扮演配位剂和稳定剂作用,也会进一步减慢反应的速率。因此反应体系中 Ag^+ 和 Cl^- 的反应速率大幅降低,导致形成四面体状 AgCl 纳米颗粒。如果反应中不添加 NaOH,如图 3.8 所示,产品形貌以立方块为主,只能产生少量的四面体。AgCl 纳米晶的形貌转化过程如图 3.9 所示。

图3.8 在二氯乙烷体系中不添加 NaOH 的情况下反应获得样品的 SEM 图片：
(a) 4 h 和 (b) 6 h

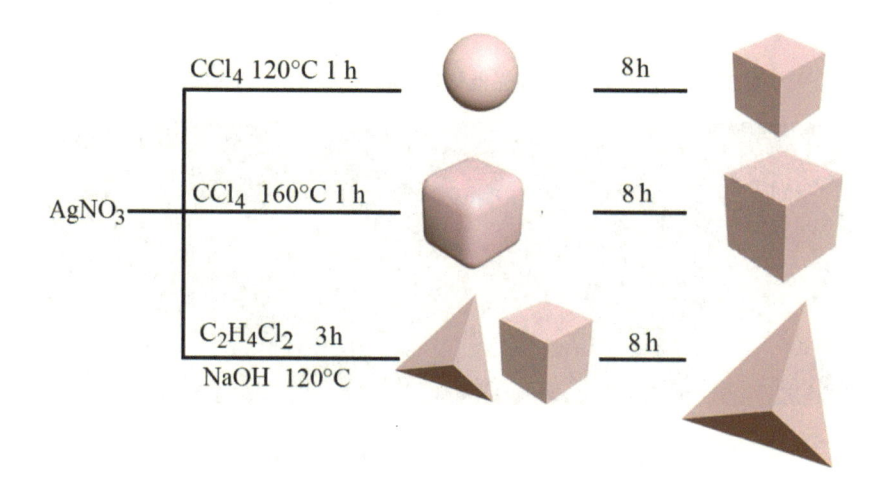

图3.9 不同形貌 AgCl 纳米晶的形成过程

AgCl 样品可在紫外灯照射下转化为 Ag/AgCl 复合光催化剂。Ag/AgCl 纳米块的 XRD 图谱（见图3.10（a））中主要衍射峰可以归属到立方相的 AgCl，而两个较弱的峰 $2\theta = 38.1°$（111）和 44.3°（200）可以归属为金属 Ag。其 SEM 图如图3.11（a）所示，从图中可以看到一些细小的 Ag 纳米颗粒附着在 AgCl 表面。

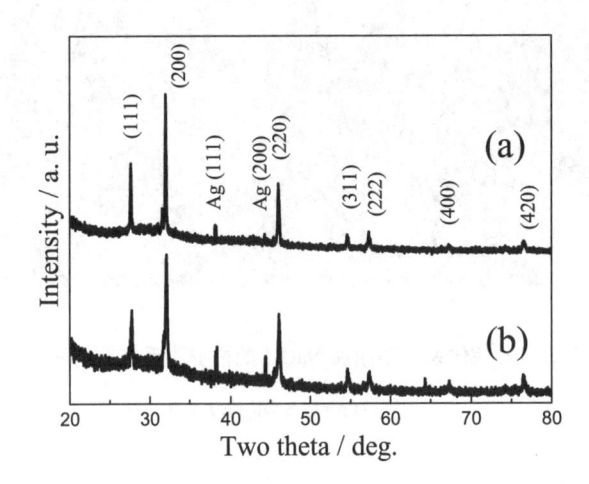

图 3.10 Ag/AgCl 纳米块光催化前（a）和后（b）的 XRD 图谱

图 3.11 Ag/AgCl 纳米块光催化前（a）和后（b）的 SEM 图

用 X-射线光电子能谱（XPS）分析了 Ag/AgCl 纳米块表面的元素状态。Ag/AgCl 纳米块的总谱显示样品含有 Ag 和 Cl 两种元素（见图 3.12（a））[23]。Cl 2p 的谱图（见图 3.12（b））中在 197.7 eV 和 199.3 eV 处有两个峰，分别对应了 Cl $2p_{3/2}$ 和 Cl $2p_{1/2}$。图 3.12（c）是 Ag 2p 的谱图，图中 368.0 eV 和 374.1 eV 的两个峰对应于 Ag $3d_{5/2}$ 和 Ag $3d_{3/2}$。对这两个峰进行了分峰处理，可在 367.9 eV、368.4 eV、374.0 eV、374.5 eV 处得到四个峰。其中 367.9 eV 和 374.0 eV 处的 XPS 峰对应于 AgCl 中的 Ag^+ 离子[180]。而 368.4 eV 和 374.5 eV 处的峰归属于 Ag^0[181-183]。通过原子组分分析可进一步得出 Ag:Ag^+ 为 10.1%。

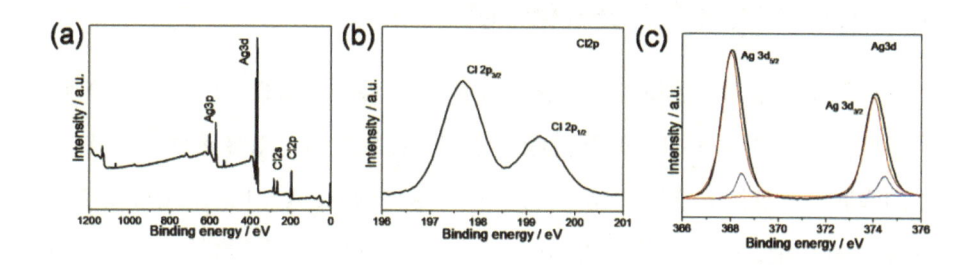

图 3.12　Ag/AgCl 纳米块的 XPS 图谱：（a）survey、（b）Cl 2p 和（c）Ag 3d

通过固体紫外-可见漫反射光谱可以测试材料的光俘获性能。从图 3-13 可以看出，AgCl 纳米块的光吸收主要在紫外区域，在可见光区的吸收很弱。Ag/AgCl 纳米块在波长 500～700 nm 的区域却有很强的吸收。这是由于 AgCl 纳米块表面的 Ag 纳米颗粒的 LSPR 效应导致了 Ag/AgCl 纳米块对可见光的吸收增强。此外，与大块的 Ag/AgCl 相比，AgCl 纳米块样品的紫外吸收边没有明显的位移，说明 AgCl 纳米块的粒径减小并未引起带宽变大或光俘获性能变差。

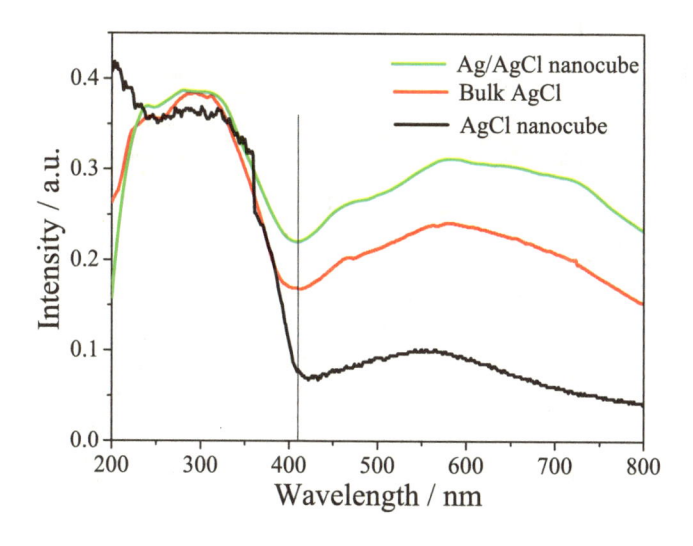

图 3.13　AgCl 纳米块、Ag/AgCl 纳米块和大块 Ag/AgCl 的漫反射紫外-可见光谱

3.3.4 Ag/AgCl 纳米晶的光催化性能

在模拟太阳光下测试了光催化剂降解 RhB 溶液的活性。图 3.14（a）给出了不同降解时间下 RhB 溶液的吸收谱图。结果显示，RhB 溶液在 554 nm 处的吸收峰强度随光照时间延长迅速下降，这说明 Ag/AgCl 纳米块可以快速降解 RhB 溶液。采用 0.1 mM AgNO$_3$ 和 0.1 mM NaCl 的溶液直接反应制备的大块 Ag/AgCl 作为对照品。如图 3.14（b）所示，光催化活性按照 Ag/AgCl 大块、Ag/AgCl 四面体、纳米球、纳米块的顺序依次升高。在模拟太阳光辐照 6 min 后，四种催化剂对 RhB 溶液的降解率分别为 20%、59%、87% 和 95%。将所得数据进行一级动力学拟合，以 ln（c/c_0）和辐照时间 t 进行拟合可得到一条直线（见图 3.14（c））。直线的斜率 k_1 和相关系数 R^2 分别列在表 3.1 中。

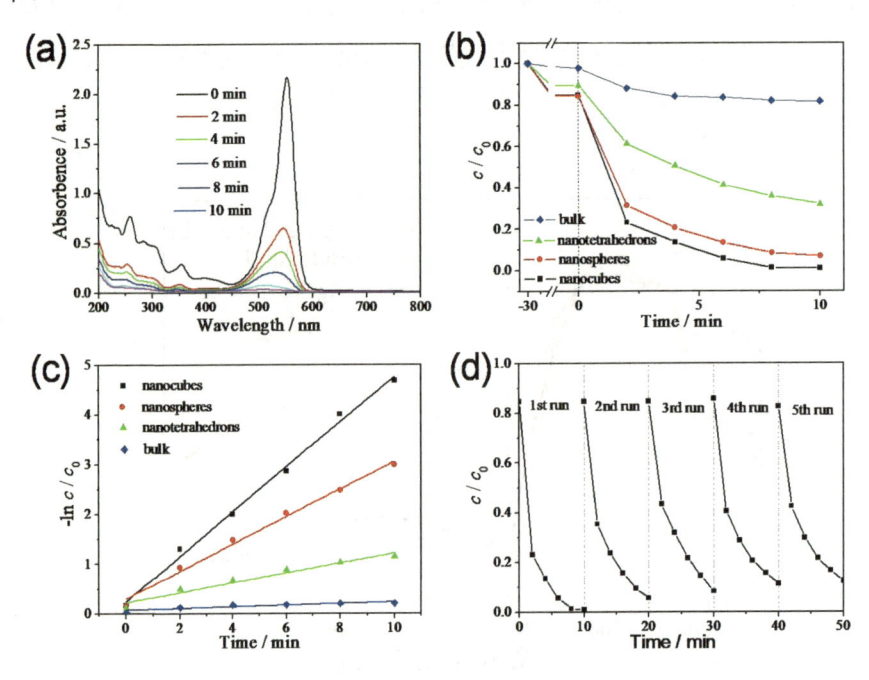

图 3.14 （a）不同反应时间下 RhB 溶液的紫外-可见光谱，（b）Ag/AgCl 降解 RhB 溶液曲线以及其（c）准一级动力学拟合曲线、（d）循环曲线

表 3.1　Ag/AgCl 样品降解 RhB 溶液的准一级动力学参数

样品	吸附量	k_1	R^2
大块	0.013	0.010	0.9189
四面体	0.107	0.099	0.9481
纳米球	0.157	0.275	0.9911
纳米块	0.153	0.451	0.9938

结果表明，Ag/AgCl 纳米块对 RhB 溶液的降解符合一级动力学，其降解速率为 0.451 min^{-1}，比大块 Ag/AgCl（0.010 min^{-1}）、四面体（0.099 min^{-1}）、纳米球（0.275 min^{-1}）都高。可能由于纳米块的粒径较小，且结晶质量高，光生电子和空穴复合概率更小，所以纳米块的催化效率高[184-185]。为了评价 Ag/AgCl 纳米块的稳定性，我们测试了其对 RhB 溶液的循环降解性能。结果如图 3.14（d）所示，经过 5 个循环后，催化剂的活性略有降低，这说明其具有较好的光稳定性。经 5 个循环后 Ag/AgCl 纳米块的 XRD 图谱（见图 3.10（b））显示 Ag^0 的峰明显增强，这说明在催化过程中产生了更多的 Ag^0。Ag/AgCl 纳米块的 SEM 图片（见图 3.11（b））也可以看出催化反应后一些纳米块的形貌被破坏。这些可能是光催化反应后 Ag/AgCl 纳米块催化性能降低的原因。

TOC 分析可以测试催化剂对 RhB 溶液的矿化率[149]。在模拟太阳光辐照 10 min 后，大块 Ag/AgCl、四面体、纳米球和纳米块的矿化率分别为 14%、57%、79% 和 83%，说明纳米块对 RhB 溶液的矿化率较高。

尽管 Ag/AgCl 催化剂在模拟太阳光下具有较高的催化活性，但是由于光源的成本较高，其应用仍受到限制。如果能在自然光下保持较高的催化活性，对于光催化剂的实际应用有重要意义。尽管实验地点选择在光照较强的内蒙古进行实验(呼和浩特，下午 1 点，晴天)，其光密度依然很低，约 5 mW·cm^{-2}。图 3.15 给出了催化剂在太阳光下对 RhB 溶液的降解效率。如图所示，100 mL

溶液中催化剂用量为 50 mg 时，经过 8 min 光催化反应对 RhB 溶液的降解率为 96%。另外，我们注意到催化剂用量越高光催化效率越好，但继续增加至 75 mg 时降解效率不再增加，这说明 50 mg 是比较合适的催化剂用量。

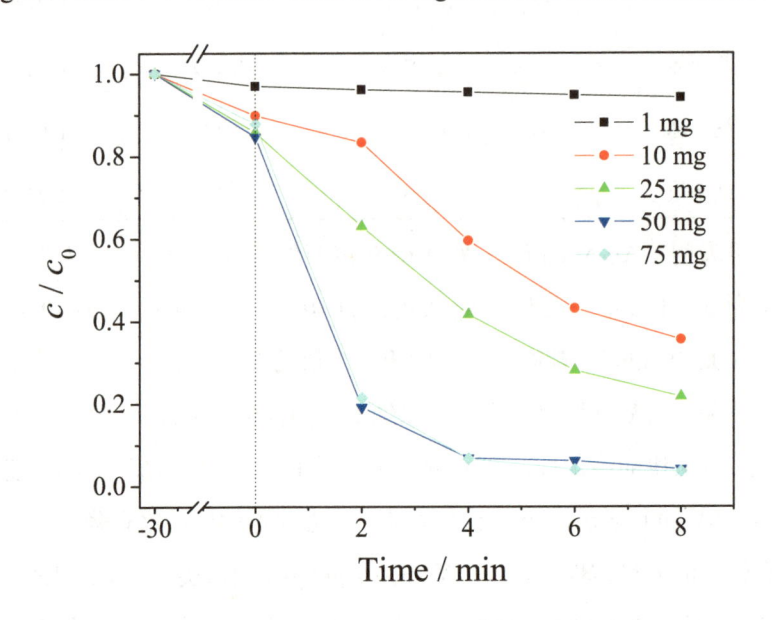

图 3.15　块状 Ag/AgCl 样品在户外阳光下对 RhB 溶液降解曲线

通常认为在光的辐照下，Ag/AgCl 会产生光生电子-空穴对（e⁻-h⁺）。光生电子和空穴转移到催化剂表面并进一步转化为羟基自由基和超氧自由基，它们被认为是光催化过程的活性物质[155]。在催化体系中加入三种物质：三乙醇胺（TEOA，空穴俘获剂）、叔丁醇（TBA，羟基自由基俘获剂）、对苯醌（BZQ，超氧自由基俘获剂），用于研究光催化过程中的主要活性物质和机理[156-157]。结果如图 3.16 所示，Ag/AgCl 纳米块在叔丁醇、对苯醌和三乙醇胺存在时的催化反应的速率常数分别为 0.389 min⁻¹、0.245 min⁻¹ 和 0.086 min⁻¹，这说明加入三乙醇胺和对苯醌明显降低了光催化的速率，而加入叔丁醇基本不影响光催化反应速率。这一结果表明，超氧自由基和空穴对光催化过程起主要作用，而羟基自由基不是 Ag/AgCl 纳米块降解 RhB 溶液的主要光催化活性物质。这一结果表明该反应的光催化机理与第 2 章的

Ag/AgCl 中空方块基本一致。

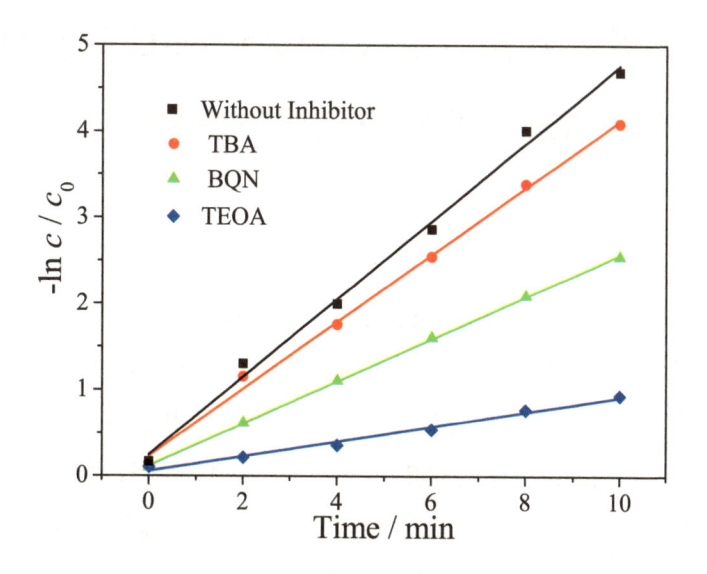

图 3.16 Ag/AgCl 纳米块在不同俘获剂添加后的准一级动力学曲线：$-\ln(c/c_0)$ - t

3.4 本章小结

本章采用一步溶剂热法制备了小粒径的 AgCl 纳米颗粒，并通过控制温度和溶剂实现了颗粒形貌控制，进而分析了 AgCl 纳米晶的生长机理。通过紫外辐照在 AgCl 纳米晶的表面原位生成了 Ag 纳米颗粒，形成 Ag/AgCl 复合光催化剂。由于粒径更小、结晶度更高，导致 Ag/AgCl 纳米块在模拟太阳光下的光催化效率很高。Ag/AgCl 纳米块在室外太阳光下同样具有很高的催化效率。机理分析结果表明，Ag/AgCl 纳米块的催化机理与第 2 章介绍的 Ag/AgCl 中空纳米块一致。这一研究为 AgCl 纳米颗粒的形貌可控制备提供了新的途径，并为 Ag/AgCl 纳米光催化剂的实际应用提供了数据支持。

第4章 多孔六氰合铁酸锌（$Zn_3[Fe(CN)_6]_2 \cdot xH_2O$）微-纳米晶的制备及吸附性能研究

4.1 引言

具有可调的微-介孔结构且比表面积大的多孔材料可以吸附许多有机化合物。除了传统的无机多孔材料如多孔硅、沸石和碳材料外[186-191]，有机-无机框架结构（MOF）也成为多孔材料研究的热点。由于这类材料具有有机和无机模块、规则的晶体结构和孔道[192]，与传统的多孔材料相比较，MOF具有一些有趣的特征，如高比表面积、可调的孔径、丰富的未饱和配位点以及吸附特定功能物质的能力[193-198]。这些性质使得MOF在气体存储、提纯、催化、传感器等领域广泛应用[199-208]。尤其是在有毒化合物（含氮或硫的化合物）吸附领域，MOF被认为是一种极具潜力的材料[209-214]。Feng等制备了阳离子有机金属铟MOF，并研究了其在阴离子交换分离过程中的应用[215]。Su等制备了一种微孔阴离子MOF，该材料可吸附镧（III）离子，并且可通过离子交换途径选择性吸附染料[216]。Ma等也报道了一种可选择性吸附有机染料的MOF[217]。因此，选择性吸附有害化合物已成为MOF基材料研究的热点[218-219]。

普鲁士蓝（PB）及其类似物（PBA）由金属与氰基桥联构成，是MOF家族重要的成员，并且在磁性和气体存储领域得到广泛研究[220-221]。近年来，各种形貌和粒径的PB或PBA微-纳米晶已被制备，并具有优异的性能[222-224]。块状和盘状的$Zn_3[Fe(CN)_6]_2 \cdot xH_2O$（ZnPBA）已经制备，并用于电池、催化剂和医药领域[225-228]。但是，其沸石样结构特征却被人们忽视[229]。与PB相比，人们很少关注锌基PBA材料的微-纳米晶[230-232]。

本章介绍一种简单的液相法制备$Zn_3[Fe(CN)_6]_2 \cdot xH_2O$（ZnPBA）微-

纳米晶的方法。通过控制反应物溶液的浓度、HCl 和聚乙烯吡咯烷酮（PVP）的用量可以调控微-纳米晶的粒径和形貌。通过 XRD、SEM、TEM、BET 等对微-纳米晶的结构和形貌进行了表征。该材料对有机染料具有选择性吸附性能，尤其对亚甲基蓝（MB）的吸附性能最佳，可以达到 1.016 $g·g^{-1}$。并且吸附的染料可以在有机溶剂中脱附，这有利于材料的重复利用。

4.2 实验部分

4.2.1 材料

本研究用到的所有化学试剂均为分析纯，购自国药集团。实验中用到的水均为去离子水。

4.2.2 制备 ZnPBA 微-纳米晶

球形 ZnPBA 微-纳米晶：典型的实验中，将 50 mL 的 $ZnCl_2$（6 mM）溶液加入 50 mL 的 $K_3Fe(CN)_6$（4 mM，含 PVP 30 mg）的溶液中。混合溶液在室温下搅拌 30 min，然后陈化 6 h。溶液离心分离后得到黄色沉淀，用去离子水和乙醇洗涤数次后在 60 °C 真空烘箱中干燥 12 h。

块状 ZnPBA 微-纳米晶：50 mL 的 $ZnCl_2$（6 mM）溶液和 1 mL HCl 溶液（12 nM）加入 50 mL 的 $K_3Fe(CN)_6$（4 mM）的溶液中。混合溶液在室温下搅拌 30 min，然后陈化 6 h。溶液离心分离后得到黄色沉淀，用去离子水和乙醇洗涤数次后在 60 °C 真空烘箱中干燥 12 h。

多面体状 ZnPBA 微-纳米晶：制备方法与块状 ZnPBA 微-纳米晶相似，只是把 HCl 用量减少到 0.1 mL。

4.2.3 仪器

物相结构由 X 射线衍射（XRD，Rigaku D/MAX 2400，Cu-K_α λ =

1.540 6 Å，扫描速率 2 °/min）表征。样品的形貌、粒径和微观结构用 FESEM（Hitachi S-4800，加速电压 20 kV）和 TEM（JEM-2100，加速电压 200 kV）表征。氮气吸附-脱附性能由 Coulter SA 3100 比表面分析仪在 77 K 下测定。测试前，样品在氮气中 373 K 下处理 12 h。紫外-可见光谱用 UV-2450 光谱仪测量。

4.2.4 染料吸附实验

染料吸附率通过测量吸附前后的染料浓度获得。本实验用亚甲基蓝（MB）、甲基橙（MO）和罗丹明 B（RhB）作为模型染料，染料的初始浓度为 20 mg·L^{-1}。吸附实验中，将 10 mg ZnPBA 样品和 500 mL 染料溶液加入一个锥形瓶。混合物超声 1 min 后进行磁力搅拌。每隔 1 h 抽取 4.0 mL 染料溶液，溶液在 10 000 rpm 下离心 5 min，去除 ZnPBA。染料溶液的剩余浓度由紫外-可见分光光度计测量，测量波长为 MB 664 nm、MO 505 nm、RhB 554 nm。吸收量（q_t, mg/g）可以由下面公式计算：

$$q_t = \frac{(c_0 - c_t)V}{m} \tag{4.1}$$

式中，c_0（mg·L^{-1}）是染料初始浓度，c_t（mg·L^{-1}）是吸附时间为 t 时的染料浓度，V（L）是染料溶液体积，m（mg）是 ZnPBA 的质量。

4.3 结果与讨论

4.3.1 制备与表征

制备的 ZnPBA 样品的晶体结构由 XRD 表征。三种形貌的样品的 XRD 图谱如图 4.1 所示，所有可检测的衍射峰均可与标准谱图中的立方相 Zn$_3$[Fe(CN)$_6$]$_2$·xH$_2$O（JCPDS 38-0687，a = 10.335 Å）相对应。这一结果证明不同条件下制备的样品均为 Zn$_3$[Fe(CN)$_6$]$_2$·xH$_2$O，并且结晶性良好。

第 4 章　多孔六氰合铁酸锌（Zn$_3$[Fe(CN)$_6$]$_2$·xH$_2$O）微-纳米晶的制备及吸附性能研究

图 4.1　不同形貌 ZnPBA 的 XRD 图谱：(a) 球形、(b) 块状和 (c) 多面体

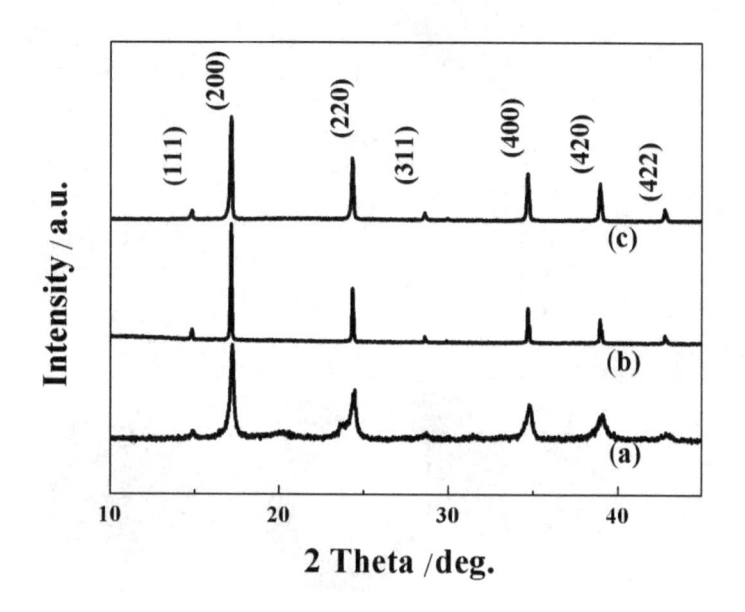

图 4.2（a）和 4.2（b）是球形 ZnPBA 的 FESEM 图。从图中可以看出，球形颗粒的直径约 300 nm，表面比较粗糙。图 4.2（c）是典型的 ZnPBA 的 TEM 图，该图进一步证明了样品具有均匀的球状形貌。当用 1 mL 盐酸代替 PVP 添加到体系中，可以获得块状 ZnPBA 微-纳米晶。块状颗粒粒径均匀，约 1 μm。颗粒的表面非常光滑，这与颗粒的结晶性好有关。当盐酸的用量减少到 0.1 mL，获得产品的形貌为多面体（见图 4.2（g）～（i）），粒径约 2 μm。从单个颗粒的放大图可以看出，多面体颗粒由 6 个正方形和 8 个六边形构成（图 4.2（h）的插图为其透视图）。上述结果表明，添加剂在 ZnPBA 微-纳米晶形貌控制中起到关键作用。

图 4.2 不同形貌 ZnPBA 的 FESEM、TEM 图：
(a)～(c) 球形颗粒、(d)～(f) 块状颗粒、(g)～(i) 多面体颗粒

4.3.2 反应条件的影响

除了本身的晶体结构特征，外部的实验条件对 ZnPBA 颗粒的粒径和形貌起到重要作用[233]。通过调节反应物和添加剂用量可实现颗粒的形貌和粒径可控制备。

实验表明，通过改变 Zn^{2+} 和 $[Fe(CN)_6]^{3-}$ 的摩尔比可以控制球形 ZnPBA 颗粒的粒径。图 4.3（a）～（d）显示了 Zn^{2+} 浓度从 4～8 mM 条件下制备的 ZnPBA 颗粒（所有实验中 Zn^{2+} 和 $[Fe(CN)_6]^{3-}$ 的摩尔比保持 3:2）。明显可以看出，颗粒的粒径随着 Zn^{2+} 浓度的增加而增加。从图中可以看出，在 Zn^{2+} 浓度较低时（4 mM 或 5 mM）获得的颗粒单分散性较好。

图 4.3（e）～（h）显示了盐酸用量对 ZnPBA 颗粒形貌和粒径的影响。以盐酸为添加剂可以获得块状和多面体状的颗粒。当盐酸的剂量从 0.8 mL

降到 0.6 mL 时，块状颗粒逐渐去角（图 4.3（e）、（f））。进一步降低盐酸的用量，从 0.4～0.2 mL，可以获得多面体状颗粒（图 4.3（g）、（h））。

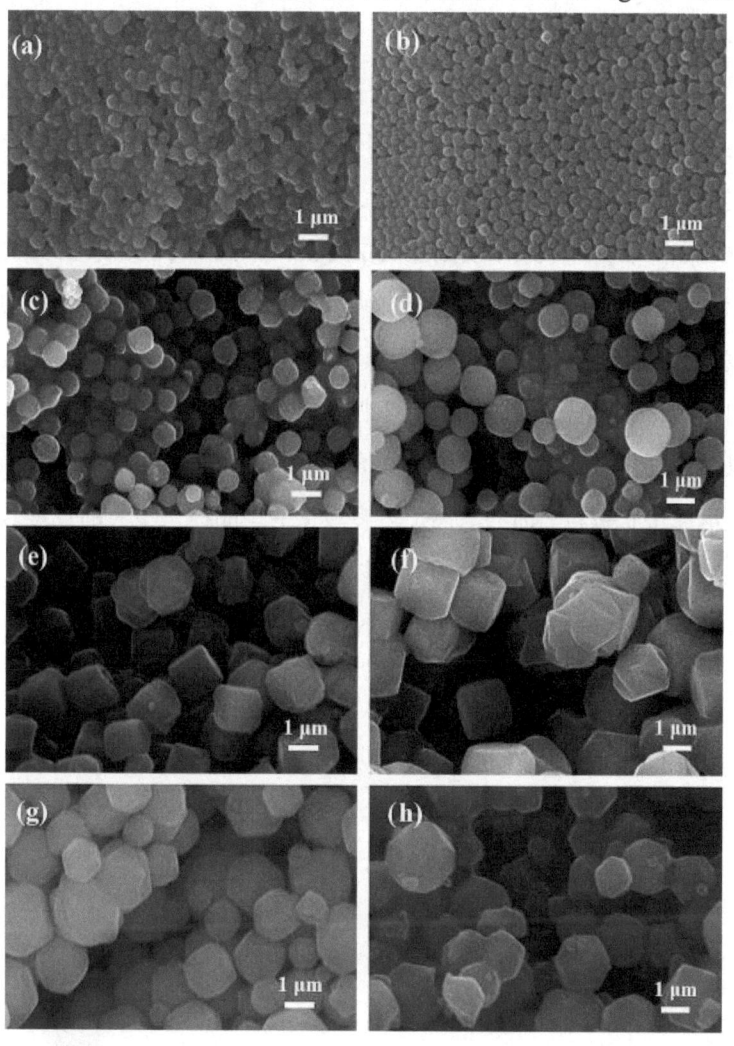

图 4.3 ZnPBA 粒径随 Zn^{2+} 浓度变化：（a）4 mM、（b）5 mM、（c）7 mM、（d）8 mM；颗粒形貌随盐酸用量减少的变化：（e）0.8 mL、（f）0.6 mL、（g）0.4 mL、（h）0.2 mL

尽管 PVP 作为添加剂控制纳米颗粒的形貌和粒径已被广泛报道，但用盐酸控制 MOF 颗粒的形貌的报道却很少。从上述实验结果可以看出，盐酸在 ZnPBA 颗粒的形成过程中发挥了重要的作用。在本实验中，$ZnCl_2$ 用作锌源，并且颗粒的形貌并未随 $ZnCl_2$ 的用量改变而改变。这表明 Cl^- 离子对颗粒形貌的影响较小，而盐酸中的氢离子（H^+）对 ZnPBA 的形貌控制起到了重要作用[234-235]。

图 4.4 给出了不同形貌 ZnPBA 微-纳米晶的形成过程。通常 MOF 颗粒的生长被描述成非经典的生长过程[236-238]。这里晶核组装和奥斯特瓦尔德熟化过程用于解释不同形貌 ZnPBA 微-纳米晶的形成机理。首先，Zn^{2+} 和 $[Fe(CN)_6]^{3-}$ 离子在体系中反应生成 ZnPBA 晶核，晶核逐渐生长成为微小的纳米晶，进而聚集形成定向生长的纳米颗粒。当 PVP 作为添加剂时，ZnPBA 晶核被 PVP 包裹，只能聚集以降低表面能。由于包覆了 PVP 聚集体是没有方向性的，最终只能形成球状的颗粒。当盐酸作为添加剂时，晶体的形成受到氢离子的影响。当氢离子浓度较高时，晶核很容易沿着高能晶面组装，并快速形成块状颗粒。而当氢离子浓度较低时，晶核自组装的速度较慢，生长速度较慢的晶面得以保存，因此可以获得多面体形的颗粒[239-241]。（111）和（100）晶面的能量和生长速率的不同可能导致了多面体颗粒的形成[242]。

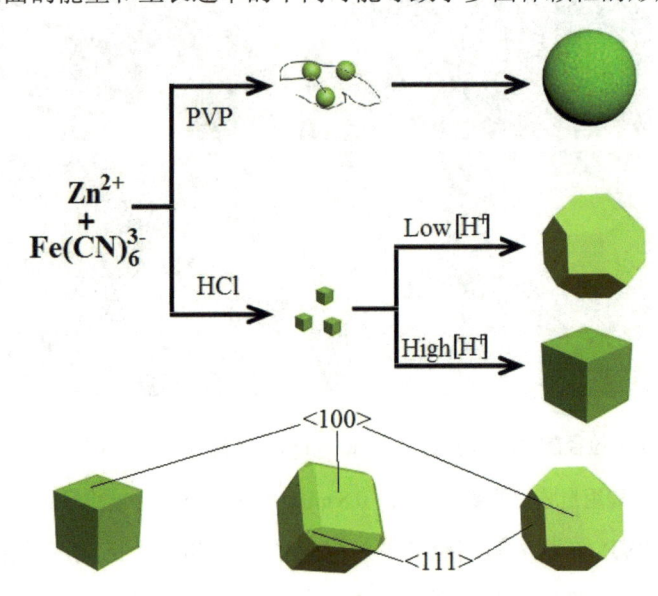

图 4.4 不同形貌的 ZnPBA 微-纳米晶的形成机理

4.3.3 ZnPBA 微-纳米晶的吸附性能

ZnPBA 微-纳米晶的多孔性能由 Brunauer-Emmett-Teller（BET）表面积测量法测量。图 4.5 是 ZnPBA 微-纳米块的氮气吸附-脱附曲线，其孔径分布图插在图 4.5 中。BET 比表面积和孔体积分别为 643.2 $m^2 \cdot g^{-1}$ 和 0.25 $cm^3 \cdot g^{-1}$。从图 4.5 中的插图可以看出，孔径范围在 1～2 nm 处有个窄峰，在 40 nm 处有一个弱的峰。

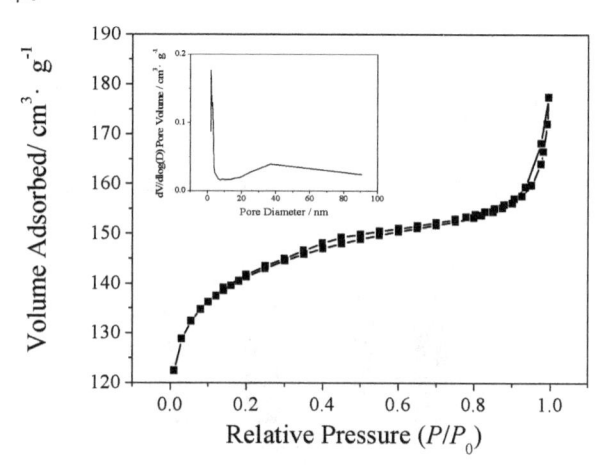

图 4.5 块状 ZnPBA 的氮气吸附-脱附图，插图为其孔尺寸分布图

考虑到 ZnPBA 的多孔性能和比表面较大，它可以用于污染物的吸附。不同形貌 ZnPBA 样品的吸附活性通过对 MB、MO、RhB 的吸附来测量。图 4.6（a）给出了 ZnPBA 块对 MB、MO、RhB 的吸附性能。从图中可以看出，对于所有染料开始的吸附速率较快，随着吸附时间的延长而逐渐减慢，10 h 后达到吸附平衡。但是三种染料的吸附量有较大差别。ZnPBA 块对 MB、MO、RhB 的最大吸附量分别是 1.016 $g \cdot g^{-1}$、0.125 $g \cdot g^{-1}$、0.010 $g \cdot g^{-1}$。这一结果表明，ZnPBA 块对 MB 具有高度选择性吸附。为验证其对 MB 的选择性吸附，将 ZnPBA 块分散在 MB/MO 或 MB/RhB 混合溶液中。如图 4.7（a）所示，MB 和 MO 混合溶液的颜色为暗紫色，加入 ZnPBA 块后溶液颜色迅速变为橙色，而黄色的粉末变为深蓝色。这表明 ZnPBA 块吸附了 MB，而

MO 留在了溶液中。同样的现象发生在 MB 和 RhB 混合溶液中，如图 4.7（b）所示。

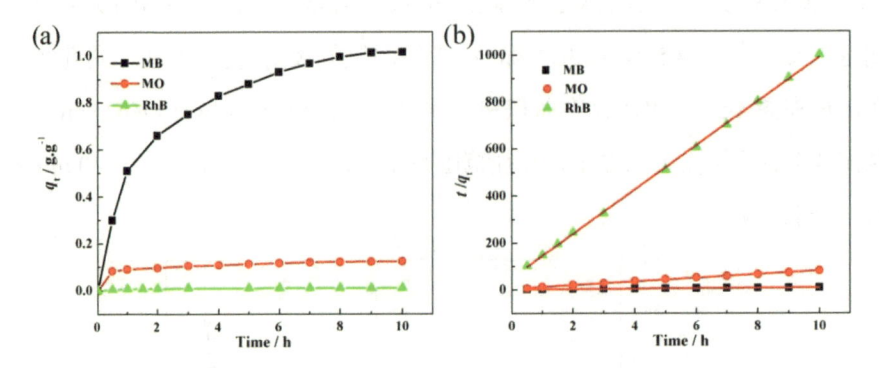

图 4.6 （a）ZnPBA 块对 MB、MO 和 RhB 的吸附曲线和（b）准二级动力学曲线

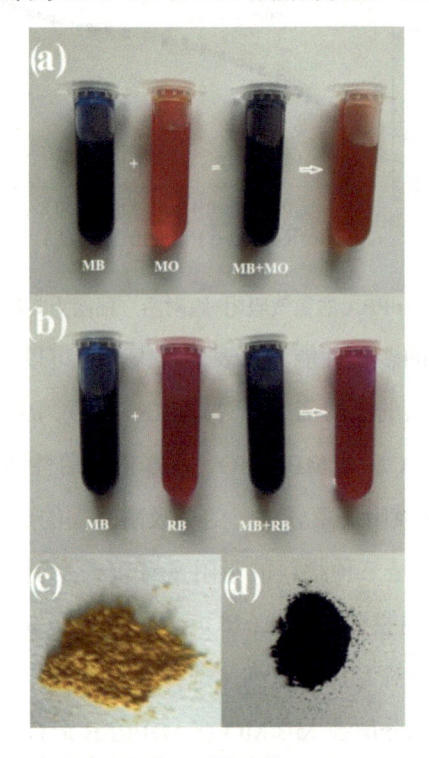

图 4.7 （a）MB、MO、混合溶液及吸附 1h 后的照片；（b）MB、RhB、混合溶液及吸附 1h 后的照片；（c）块状 ZnPBA 粉末的照片；（d）吸附 MB 后块状 ZnPBA 粉末的照片

球形和多面体形的 ZnPBA 颗粒对染料的吸附性能列在表 4.1 中。从表中可以看出,大块的 ZnPBA 也可以吸附染料,但是其吸附能力较微-纳米结构颗粒低,这可能是由于大块颗粒的表面积较小。值得注意的是,球形的颗粒尽管粒径较小,但其吸附能力比块状和多面体状颗粒低。这可能是由于球状颗粒的一些活性位点被 PVP 占据。

表 4.1 不同形貌的 ZnPBA 样品的吸附能力

形貌	MB/(g·g^{-1})	MO/(g·g^{-1})	RhB/(g·g^{-1})
块状	1.016	0.125	0.010
球状	0.731	0.085	0.008
多面体	0.935	0.127	0.009
大块	0.702	0.081	0.007

为了考察吸附动力学,准一级动力学(4.2)和准二级动力学(4.3)模型用于处理吸附数据[243]。公式如下:

$$\ln(q_e - q_t) = \ln q_e - k_1 t \tag{4.2}$$

$$\frac{t}{q_t} = \frac{1}{k_2 q_e^2} + \frac{t}{q_e} \tag{4.3}$$

式中,q_e 和 q_t(g·g^{-1})分别是吸附平衡时和时间 t 时的吸附量,k_1(h^{-1})是准一级动力学常数,k_2(g·g^{-1}·h^{-1})是准二级动力学常数。ln(q_e-q_t)-t 或 t/q_t-t,如果为直线,表明该动力学模型比较合适。拟合结果显示二级动力学模型与本实验结果更加接近。另外反应速率 v_0(g·g^{-1}·h^{-1})可以通过下面公式计算:

$$v_0 = k_2 q_e^2 \tag{4.4}$$

所有的动力学参数和方差 R^2 均列在表 4.2 中。从表中可以看出,所有的曲线线性关系较好($R^2>0.995$),并且实验得到的最大吸附量与理论值比较吻合，说明染料的吸附过程符合准二级动力学控制。另外，该材料对 MB 的吸附性能与 MIL-100（Fe）相当[244]，比表 4.3 中列出的其他材料吸附性能好。

表 4.2 块状 ZnPBA 颗粒对染料吸附的准二级动力学参数

染料	$q_{e,exp}$/ (g·g^{-1})	$q_{e,cal}$/ (g·g^{-1})	k_2/ (g·g^{-1}·h^{-1})	R^2	V_0/ (g·g^{-1}·h^{-1})
MB	1.016	1.169	0.54	0.9977	0.739
MO	0.125	0.129	13.71	0.9975	0.228
RhB	0.010	0.011	164.96	0.9994	0.020

表 4.3 文献中报道的一些 MOF 材料对 MB 的最大吸附量

材料	MB 吸附量/（g·g^{-1}）	文献
MOF-235	0.447	[245]
MIL-100（Fe）	1.105	[244]
HKUST-1/GO	0.182	[246]
Double-Shelled C/SiO$_2$	0.171	[247]
POM@MIL-101	0.371	[248]
Activated carbons	0.344	[249]
Fe$_3$O$_4$/Cu$_3$（BTC）$_2$	0.245	[250]

对比样品吸附 MB 前后的 XRD 图谱变化可对吸附机理进行解释。如图 4.8 所示，样品吸附染料后出现了 Zn$_3$[Fe(CN)$_6$]$_2$·xH$_2$O 和 Zn$_3$[Fe(CN)$_6$]$_2$ 两组衍射峰，并且随着染料吸附量的增加，Zn$_3$[Fe(CN)$_6$]$_2$ 的衍射峰逐渐增强（见图 4.8（a）～（c））。这说明染料分子替换了 ZnPBA 中的水分子，同时使晶格参数从 10.33 Å 变为 10.95 Å。有趣的是，当染料脱附以后，晶体结构又

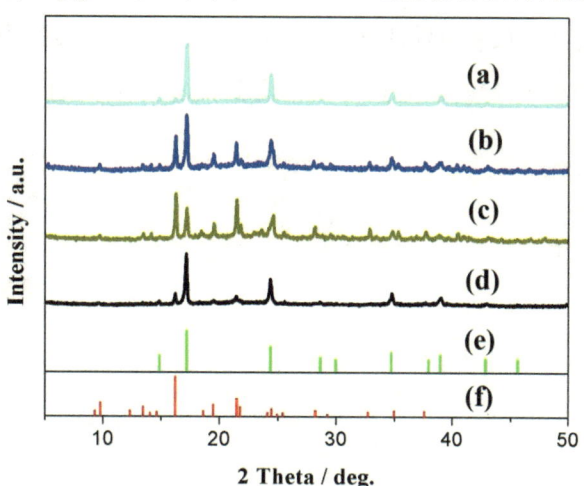

第 4 章　多孔六氰合铁酸锌（Zn$_3$[Fe(CN)$_6$]$_2$·xH$_2$O）微-纳米晶的制备及吸附性能研究

恢复为 Zn$_3$[Fe(CN)$_6$]$_2$·xH$_2$O（见图 4.8（d））。这也说明该材料的稳定性较好。

图 4.8　块状 ZnPBA 样品吸附不同染料后的 XRD 图谱：（a）吸附 RhB、（b）吸附 MO、（c）吸附 MB、（d）MB 脱附后；标准 XRD 图谱（e）Zn$_3$[Fe(CN)$_6$]$_2$·xH2O、（f）Zn$_3$[Fe(CN)$_6$]$_2$

上述实验表明 ZnPBA 微-纳米晶对 MB 具有选择性吸附。ZnPBA 表现为多孔结构，并且在晶格内存在结晶水。通常认为六氰合金属位点处的孔道对吸附起到重要的作用。平常 ZnPBA 中的金属位点由水分子占据，当较强的路易斯碱进入位点，水分子会被取代[251]。三种有机染料的分子结构在图 4.9 中给出。ZnPBA 中的孔道（10.33×10.33 Å）对 MB 分子中的—N(CH$_3$)$_2$ 和—C$_6$H$_5$ 基团而言足够大。而 MB 中的路易斯碱—N(CH$_3$)$_2$ 和 ZnPBA 中的路易斯酸-铁位点有强相互作用，这导致水分子被 MB 分子取代。结果 MB 分子可以进入 ZnPBA 的孔道产生吸附。除了路易斯酸碱的影响，染料分子的大小也对吸附产生影响。如果染料分子的尺寸比 ZnPBA 的孔径小，就可以进入 ZnPBA，否则不能进入[235]。MB 分子能够进入是因为该分子有两个短轴分别为 4.59 Å 和 8.01 Å，一个长轴为 16.75 Å。而对于 RhB（15.9×11.8×5.6 Å）而言，分子过大而不能进入 ZnPBA 的孔道，只能在表面吸附。尽管 MO（5.31×7.25×17.39 Å）的分子也可进入孔道，但由于分子较大，其吸附量小于 MB。事实上，阳离子染料的选择性吸附与电性相互作用的关系已有相关报道，如 [(C$_2$H$_5$)$_2$NH$_2$]$_2$[Mn$_6$(L)(OH)$_2$(H$_2$O)$_6$]·4DEF [223] 和 MOF-235 [245]。据此可以认为，该材料对 MB 的选择性吸附受到尺寸和电荷的双重影响。

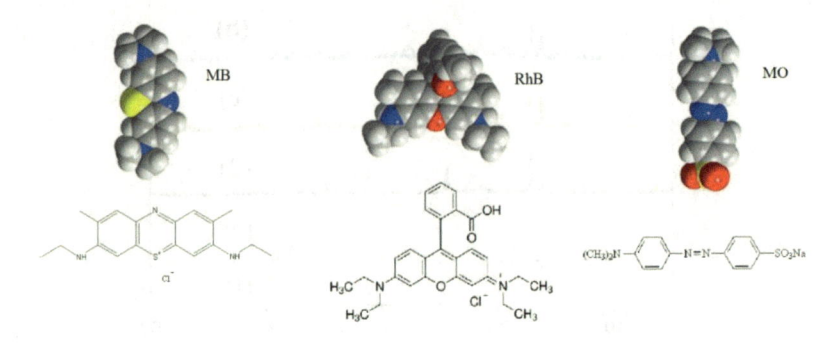

图 4.9　三种有机染料 MB、MO 和 RhB 的分子结构图[211, 213]

吸附的染料可以通过索氏提取法进行脱附。100 mg 吸附了 MB 的 ZnPBA 样品置于索式提取器中，在 500 mL 的圆底烧瓶中加入 200 mL 乙醇进行提取。经过 4 h 的提取，样品粉末由深蓝色变为黄色。这说明染料分子可以被完全洗掉。脱附的粉末可以重复利用，如图 4.10 所示，经过 5 个循环后，吸附能力变化较小，这说明该材料可以用于有机染料废水的回收利用。

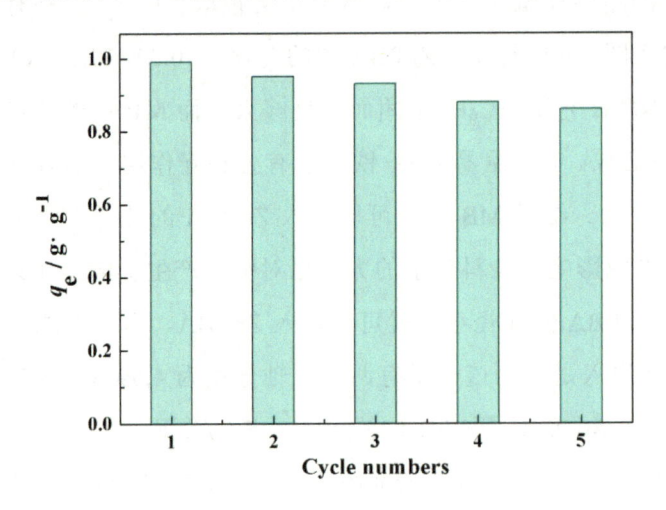

图 4.10　ZnPBA 样品对 MB 的 5 个吸附-脱附循环

4.4 本章小结

本章利用简单的室温溶液法制备了球形、块状、多面体形的 ZnPBA 微-纳米晶。并且发现了改变盐酸用量可以有效地调节 ZnPBA 的形貌。制备的 ZnPBA 微-纳米晶可以从有机染料废水中有效地分离 MB。对于 PBA 的形貌可控制备和应用而言，这项工作是一次新的尝试。

第5章 多孔/中空 Ag/ZnO/ZnFe$_2$O$_4$ 三元复合物的制备及其光催化性能研究

5.1 引言

铁酸锌具有窄的带宽 1.9 eV，这使它对太阳光的利用度提高，从而具有可见光催化性能，并且具有光稳定性、低毒性和磁性[252-254]。纳米结构的铁酸锌颗粒由于粒径较小、表面积较大，光生电子和空穴的传递距离使其更容易传递到催化剂表面，增加了光催化的表面位点，因此其在光催化领域备受关注[255-257]。但是纳米颗粒中的缺陷会增加，这些缺陷会成为光生电子和空穴的复合中心，并导致单一铁酸锌纳米颗粒的光催化性能很低。通常 p 型半导体和 n 型半导体材料会形成 II 型价带匹配，这一结构导致光生电子和空穴分离，降低了光生电子和空穴复合的概率[258-263]。最近 Guo 等报道了将 p 型 ZnFe$_2$O$_4$ 和 n 型 ZnO 用于复合提高光催化活性[264]。此外，在氧化物半导体材料中掺杂贵金属（例如银）作为共催化剂同样可以抑制光生电子和空穴的分离[265]。例如，二元复合物 Ag/ZnFe$_2$O$_4$ 和 Ag/ZnO 的光催化性能均较其单一成分的 ZnFe$_2$O$_4$ 和 ZnO 高[252, 266]。如果用 Ag 和 ZnO 修饰 ZnFe$_2$O$_4$ 形成三元复合物将更大程度地提高光催化性能。但是由于制备难度较大，到目前为止 Ag/ZnO/ZnFe$_2$O$_4$ 三元复合物还未见报道。

MOF 是含有金属离子和有机配体的多孔材料，它们在吸附、气体存储、催化领域有着广泛的应用。近期通过简单的煅烧，一些简单的 MOF 作为前驱体被用于制备金属氧化物纳米颗粒[267-270]。煅烧过程中可以通过控制温度和加热程序调控纳米颗粒的微观结构和孔洞尺寸[271-272]。作为 MOF 家族的重要成员，普鲁士蓝及其类似物已被用于制备单金属或双金属氧化物的有效前驱体[273-278]。例如，Lou 的课题组用普鲁士蓝的微-纳米块烧出了 Fe_2O_3 方盒结构，该结构具有优异的锂存储性能[276]。Chen 等通过煅烧 $M_3^{II}[Co(CN)_6]_2 \cdot xH_2O$（M = Co，Mn，Zn）获得了一系列的多孔金属氧化物，如 Co_3O_4 纳米笼[277]、$Mn_xCo_{3-x}O_4$ 纳米块[274]和 ZnO/Co_3O_4 多孔球[278]。Jiang 等通过煅烧 $K_2Zn_3[Fe(CN)_6]_2 \cdot xH_2O$ 微板获得了多孔的 $ZnFe_{2-x}O_4$-ZnO 二元复合物。但是由于普鲁士蓝类似物成分的限制，通过煅烧获得三元或三元以上的复合物还有一定难度。

本章通过煅烧吸附了银离子的普鲁士蓝类似物 $Zn_3[Fe(CN)_6]_2 \cdot xH_2O$（ZnPBA）获得了高度均匀的中空或多孔结构 $Ag/ZnO/ZnFe_2O_4$ 三元复合催化剂。由于 ZnPBA 可以吸附银离子，因此三元复合物 $Ag/ZnO/ZnFe_2O_4$ 中银的含量可以通过控制银离子在前驱体中的吸附量来调节。该催化剂对亚甲基蓝（MB）具有很好的降解能力。II 型的半导体价带匹配（$ZnO/ZnFe_2O_4$）和肖特基势垒（Ag/ZnO）对催化剂性能提高起到关键作用。

5.2 实验部分

5.2.1 材料

本实验用到的所有化学试剂均为分析纯，购自国药集团。实验中用到的水均为去离子水。

5.2.2 材料的制备

$Zn_3[Fe(CN)_6]_2 \cdot xH_2O$（ZnPBA）纳米块的制备：将 50 mL 的 $ZnCl_2$ 溶液（6 mM）和 1.0 mL 的盐酸溶液（12M）混合均匀后加入 50 mL 的六氰合铁酸钾溶液（4 mM）中。混合溶液在室温下搅拌 30 min，然后再暗处陈化 6 h。获得的沉淀通过离心分离用水和乙醇分别洗涤数次，之后在 60 °C 烘箱中干燥 12 h。

Ag 掺杂的 ZnPBA（Ag-ZnPBA）纳米块制备：将 1.5 mL 的 AgNO$_3$ 溶液（1 mg·mL^{-1}）加入 ZnPBA（100 mg 分散在 50 mL 水中）纳米块的溶液中。搅拌 2 h 后离心分离，用蒸馏水洗涤数次，之后在 60 °C 烘箱中干燥 12 h。

Ag-ZnPBA 转化为多孔的 Ag/ZnO/ZnFe$_2$O$_4$ 纳米块：将 Ag-ZnPBA 纳米块在马弗炉中以 10 °C·min^{-1} 的速度加热到 500 °C，保持温度 4 h，然后自然降温到室温，收集样品。该样品标记为 *p*-Ag/ZnO/ZnFe$_2$O$_4$。

Ag-ZnPBA 转化为中空的 Ag/ZnO/ZnFe$_2$O$_4$ 纳米块：将 Ag-ZnPBA 纳米块在马弗炉中以 2 °C·min^{-1} 的速度加热到 200 °C，保持温度 1 h。接着以 1 °C·min^{-1} 的速度加热到 500 °C，保持温度 4 h，然后自然降温到室温，收集样品。该样品标记为 *h*-Ag/ZnO/ZnFe$_2$O$_4$。

5.2.3 仪器

物相结构由 X 射线衍射（XRD，Rigaku D/MAX 2400，Cu-K_α λ = 1.540 6 Å，扫描速率 2 °/min）表征。样品的形貌、粒径和微观结构用 FESEM（Hitachi S-4800，加速电压 20 kV）和 TEM（JEM-2100，加速电压 200 kV）表征。成分测试通过能谱测试（EDX）表征。氮气吸附实验在 77 K 下获得（Coulter SA 3100 表面积分析仪）。元素 Mapping 图像在 Tecnai G2 F30 S-Twin TEM 上获得。紫外可见光谱均由 UV-2450 紫外-可见分光光度计检测。

5.2.4 光催化性能

样品的光催化性能用在模拟太阳光下降解亚甲基蓝（MB）的方法来测试。250 W 的氙灯置于石英套管反应器 10 cm 处。套管夹层中通冷凝水，用于降低氙灯辐射产生的热量。将 50 mg 的光催化剂加入 100 mL 的 MB 溶液中，MB 溶液的浓度为 10 mg·L^{-1}。在辐照前样品在避光条件下搅拌 30 min，用于使催化剂与染料达到吸附平衡。吸附平衡后开始光催化反应。到达设定的反应时间后抽取 4 mL 悬浊液，离心分离去除催化剂。MB 的浓度用紫外-

可见分光光度计在 664 nm 处测量。总有机碳（TOC）用 TOC 分析仪测定（Analytik Jena AG multi N/C 2100）。

5.2.5 光电性能

光电性能用 CHI750D 电化学工作站（上海辰华）通过标准的三电极系统测量，光源为 500 W 氙灯。典型的实验中，在 15 mg 制备的样品中加入 5 mL 乙醇、0.05 mL 油酸和 1 mg 聚乙烯吡咯烷酮（PVP），磁力搅拌 24 h。样品滴到 1 cm × 2 cm 的 FTO 玻璃上，并在 45 °C 下干燥。干燥后在 500 °C 下处理 30 min 后去掉油酸和 PVP。将覆盖有样品的 FTO 玻璃作为工作电极进行光电性能测试。铂电极和饱和甘汞电极分别作为对电极和参比电极。0.5 M 的 Na_2SO_4 溶液作为电解质。

5.3 结果与讨论

5.3.1 制备与表征

前驱体 ZnPBA 和 Ag-ZnPBA 的 XRD 谱图如图 5.1(a)所示，图中 ZnPBA 所有的峰均与面心立方的 $Zn_3[Fe(CN)_6]_2 \cdot xH_2O$（JCPDS 38-0687）的标准图谱一致，未见杂质。而在 Ag-ZnPBA 的 XRD 谱图中，除了有 $Zn_3[Fe(CN)_6]_2 \cdot xH_2O$ 的衍射峰，还包含 $Zn_3[Fe(CN)_6]_2$（JCPDS 25-1022）的峰。这一结果说明，银离子可能取代了 $Zn_3[Fe(CN)_6]_2 \cdot xH_2O$ 晶格中的部分水分子，这与已报道的 MOFs 吸附重金属离子如 Pb^{2+}、Cu^{2+}、Ni^{2+} 的结果一致[279-280]。图 5.1（b）为样品 ZnPBA 的 SEM 图，图中可以看出 ZnPBA 为规则的立方块状，边长约 1 μm。当 ZnPBA 吸附银离子后，Ag-ZnPBA 的形貌未发生明显改变，其结果如图 5.1（c）和 5.1（d）所示。并且颗粒的表面比较光滑。

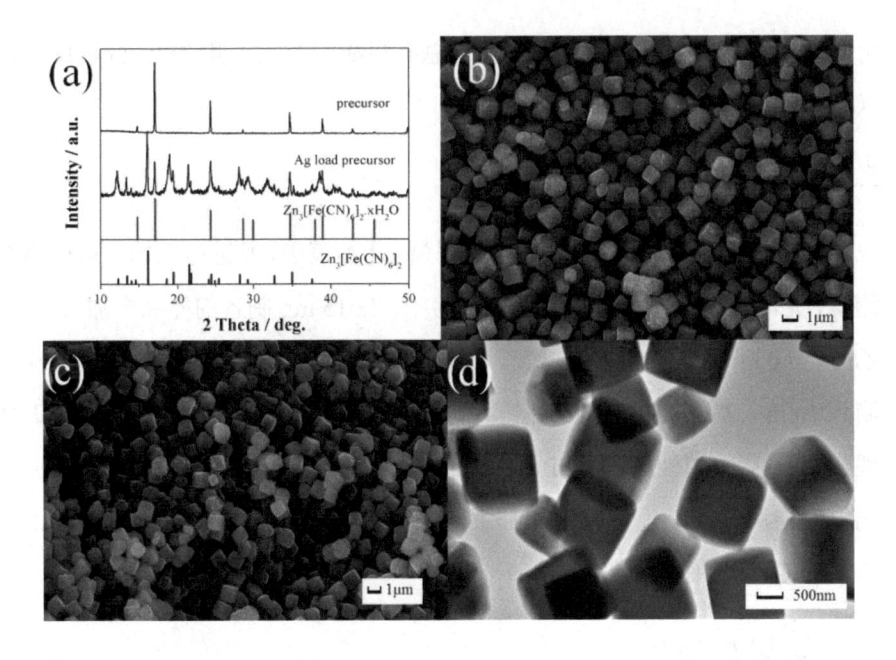

图5.1 （a）ZnPBA 和 Ag-ZnPBA 的 XRD 图谱；（b）ZnPBA 的 SEM 图；（c）Ag-ZnPBA 的 SEM 图和（d）TEM 图

制备的 Ag-ZnPBA 前驱体在空气中煅烧可以转化为多孔的或中空的 Ag/ZnO/ZnFe$_2$O$_4$ 立方块。最终的产品的形貌和结构由 FESEM 和 TEM 检测。如图 5.2（a）和 5.2（b）所示，p-Ag/ZnO/ZnFe$_2$O$_4$ 和 h-Ag/ZnO/ZnFe$_2$O$_4$ 的形貌与前驱体 Ag-ZnPBA 一致，均为块状。但是颗粒的表面变得粗糙，并且有部分颗粒破裂。图 5.2（c）是 p-Ag/ZnO/ZnFe$_2$O$_4$ 的 TEM 谱图，如图所示，经过一步快速加热以后获得的颗粒呈现多孔结构，多孔颗粒的尺寸约为 1 μm。通过两步缓慢的加热反应，可以获得中空的 h-Ag/ZnO/ZnFe$_2$O$_4$ 立方块（见图 5.2（d））。中空颗粒的尺寸约为 1 μm，壳的厚度约 20 nm。从 p-Ag/ZnO/ZnFe$_2$O$_4$（见图 5.2（e））和 h-Ag/ZnO/ZnFe$_2$O$_4$（见图 5.2（f））的 HRTEM 图可以看出，两个颗粒均呈现清晰的晶格条纹。晶格间距为 0.487 nm、0.26 nm 和 0.23 nm 的晶面分别对应立方 ZnFe$_2$O$_4$ 的（111）面、六方 ZnO 的（0002）面和立方 Ag 的（111）面。这说明样品颗粒中包含了 ZnFe$_2$O$_4$、ZnO

和 Ag 纳米颗粒。图 5.2（e）和 5.2（f）中的插图为样品的快速傅里叶变换（FFT）图谱，进一步证明了颗粒中含有 $ZnFe_2O_4$、ZnO 和 Ag。

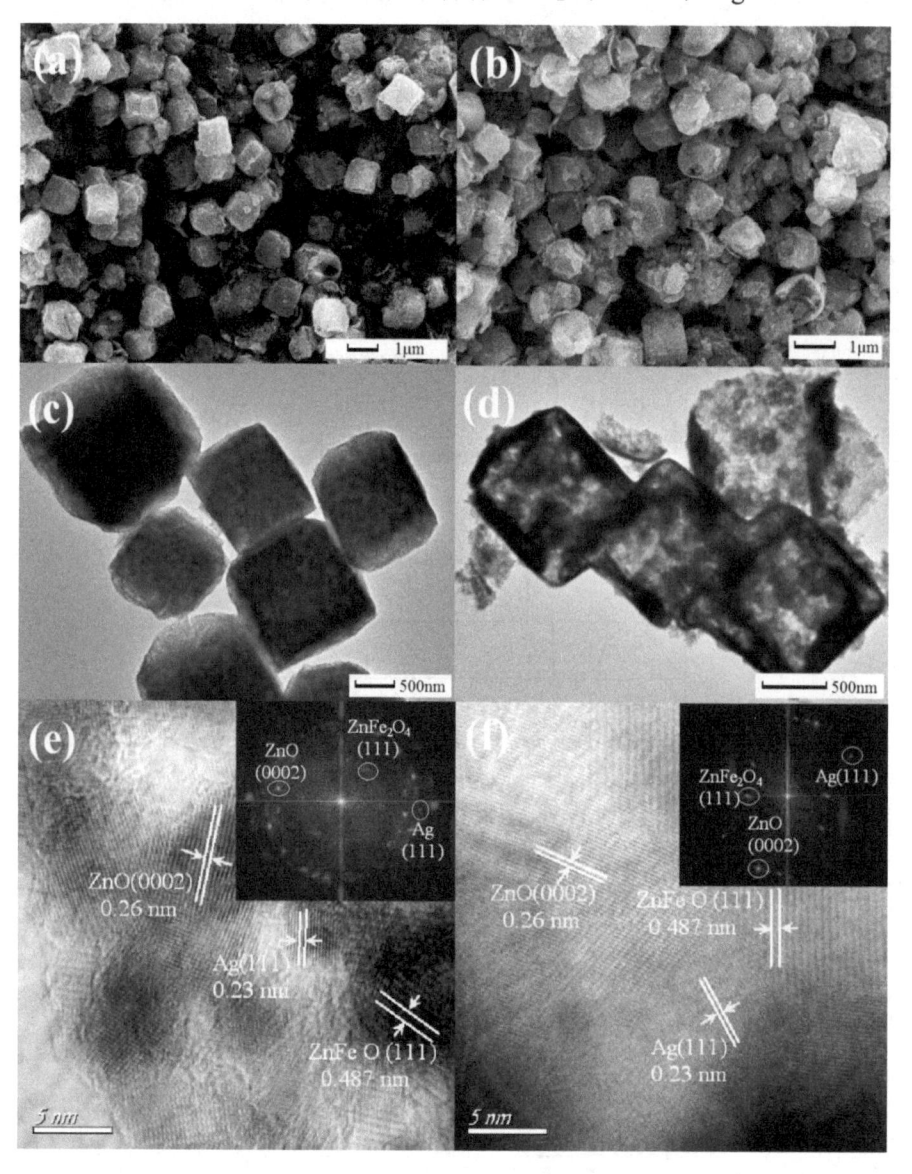

图 5.2 （a，c，e）p-Ag/ZnO/$ZnFe_2O_4$ 和（b，d，f）h-Ag/ZnO/$ZnFe_2O_4$ 的 FESEM、TEM 和 HRTEM 图；（e）和（f）中的插图为其 FFT 图

XRD 结果可以进一步证明，Ag-ZnPBA 前驱体可以通过煅烧转化成 Ag/ZnO/ZnFe$_2$O$_4$ 复合物。图 5.3 给出了 p-Ag/ZnO/ZnFe$_2$O$_4$ 和 h-Ag/ZnO/ZnFe$_2$O$_4$ 的 XRD 图谱。图谱中的所有衍射峰均可归属于三种成分：图中用"◇"标记的峰对应于六方的 ZnO（JCPDS No. 36-1451），而"○"标记对应于立方结构的 ZnFe$_2$O$_4$（JCPDS No. 22-1012），"△"标记对应于面心立方的 Ag（JCPDS No. 65-2871）。这一结果表明，p-Ag/ZnO/ZnFe$_2$O$_4$ 和 h-Ag/ZnO/ZnFe$_2$O$_4$ 中均包含 ZnFe$_2$O$_4$、ZnO 和 Ag 三种成分。

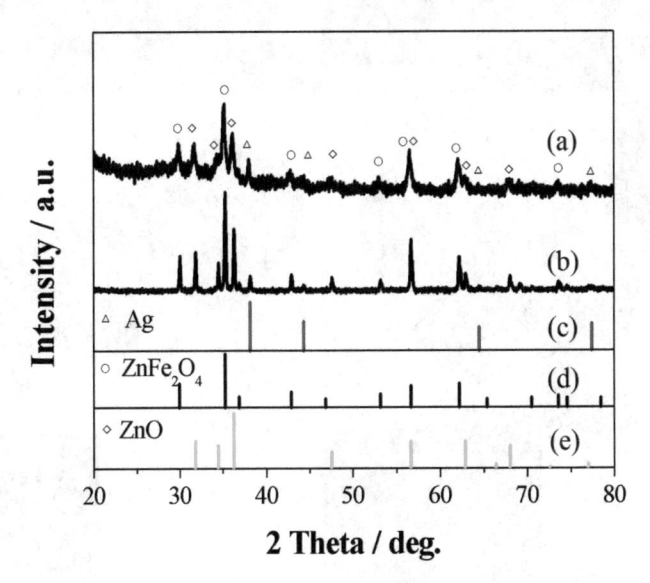

图 5.3 （a）p-Ag/ZnO/ZnFe$_2$O$_4$ 和（b）h-Ag/ZnO/ZnFe$_2$O$_4$ 的 XRD 谱图；（c）Ag、（d）ZnO 和（e）ZnFe$_2$O$_4$ 的标准谱图

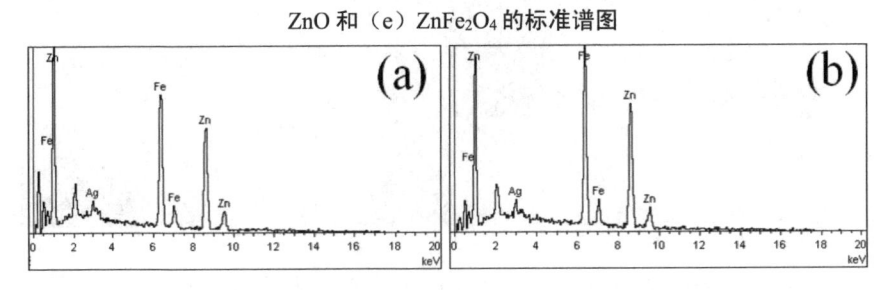

图 5.4 （a）p-Ag/ZnO/ZnFe$_2$O$_4$ 和（b）h-Ag/ZnO/ZnFe$_2$O$_4$ 的 EDX 图谱

第 5 章 多孔/中空 Ag/ZnO/ZnFe$_2$O$_4$ 三元复合物的制备及其光催化性能研究

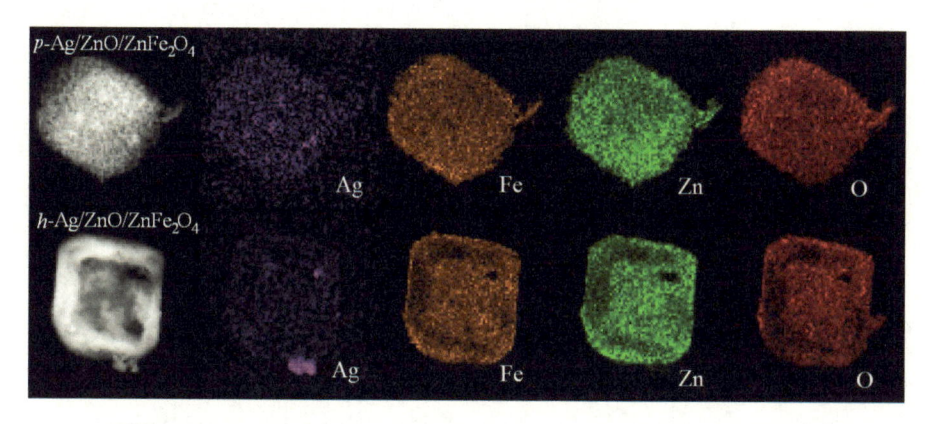

p-Ag/ZnO/ZnFe$_2$O$_4$ 和 h-Ag/ZnO/ZnFe$_2$O$_4$ 的 EDX 图谱（见图 5.4）表明样品中均存在 Zn、Fe、Ag 和 O 元素，这一结果进一步证明样品中存在 Ag、ZnO 和 ZnFe$_2$O$_4$ 三种成分。银元素在两种形貌的样品中含量相当，质量分数均为 2.17%。样品中的 Zn、Fe、Ag 和 O 元素均进行了元素 Mapping 测试，其结果如图 5.5 所示。这一结果表明通过加热掺杂的 MOFs 前驱体可以获得成分均匀的三元复合物 Ag/ZnO/ZnFe$_2$O$_4$。

图 5.5 p-Ag/ZnO/ZnFe$_2$O$_4$ 和 h-Ag/ZnO/ZnFe$_2$O$_4$ 的元素 Mapping 图谱

对 p-Ag/ZnO/ZnFe$_2$O$_4$ 和 h-Ag/ZnO/ZnFe$_2$O$_4$ 样品进行了 N$_2$ 吸附-脱附实验。图 5.6 给出了样品的 N$_2$ 吸附-脱附曲线和孔尺寸分布。结果表明 p-Ag/ZnO/ZnFe$_2$O$_4$ 和 h-Ag/ZnO/ZnFe$_2$O$_4$ 样品的 BET 表面积分别为 38.29 m$^2 \cdot$g^{-1} 和 63.51 m$^2 \cdot$g^{-1}。孔尺寸通过 Barrett-Joyner-Halenda（BJH）法获得，结果显示多孔颗粒 p-Ag/ZnO/ZnFe$_2$O$_4$ 的孔径分布在 30 nm 附近有较宽的峰，而中空样品 h-Ag/ZnO/ZnFe$_2$O$_4$ 在 8.3 nm 和 25 nm 处有两个峰。在前驱体的转化过程中，C 和 N 被氧化成气体并逸出，导致颗粒内部形成约 8 nm 的孔洞。一些孔洞彼此连接形成更大的孔洞，这导致孔径分布较宽[281-282]。结果表明，中空的颗粒较多孔颗粒具有更大的比表面积和更小的孔径，这导致其吸附和催化性能更好。

图 5.6 （a）p-Ag/ZnO/ZnFe$_2$O$_4$ 和（b）h-Ag/ZnO/ZnFe$_2$O$_4$ 的 N$_2$ 吸附-脱附分析，两图中的插图为其 BJH 孔径分布图

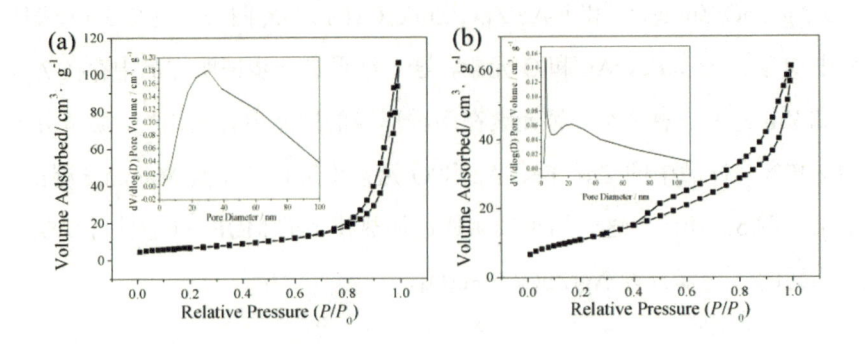

上述结果表明,通过简单的煅烧过程可以将 Ag-ZnPBA 前驱体转化为三元复合物 Ag/ZnO/ZnFe$_2$O$_4$。并且三元复合物的结构可以通过改变加热方式调节为多孔或者中空。如图 5.7 所示，多孔块可通过一步快速加热的方法获得，而中空块可以通过两步的慢速加热途径获得。当加热速率较快时，样品快速释放 CO$_2$、N$_2$ 和 H$_2$O，导致颗粒向内部塌陷，形成多孔结构。如果在 200 ℃下加热一段时间，前驱体中的 H$_2$O 分子先逸出，形成一些通道。在进一步的加热过程中，空气容易向颗粒内部扩散，在颗粒的内部更容易发生氧化，在气体逸出的同时内部物质也向外扩散，导致形成了中空结构。通过相同的两步加热程序，我们还获得了 ZnO/ZnFe$_2$O$_4$ 的二元复合物中空结构。其 XRD、FESEM、TEM 和 HRTEM 如图 5.8 所示。

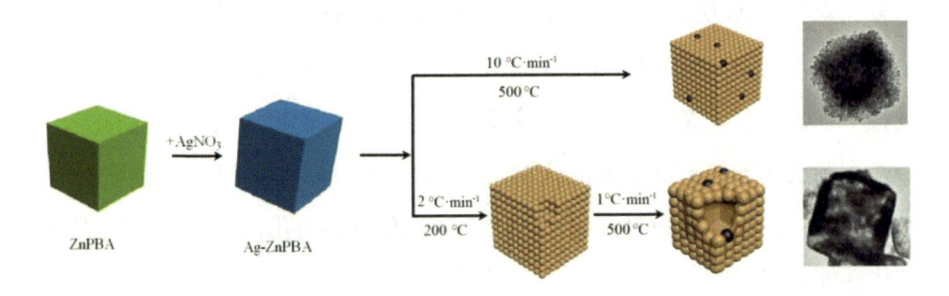

图 5.7 中空/多孔 Ag/ZnO/ZnFe$_2$O$_4$ 的形成机理

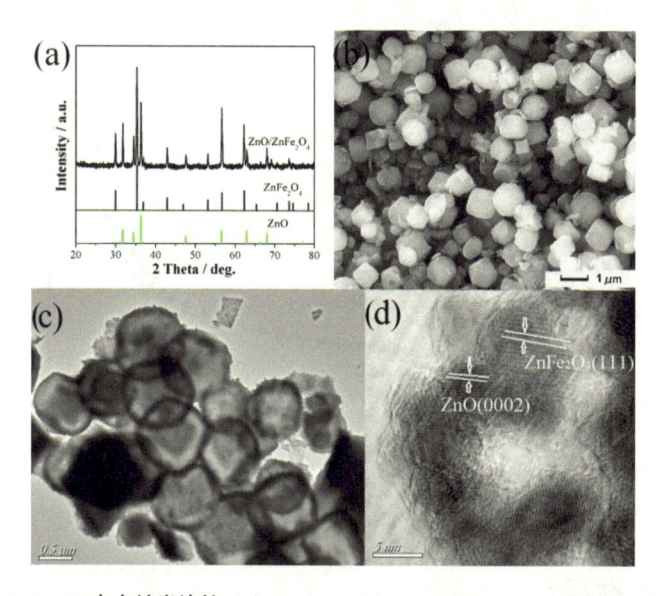

图 5.8 ZnO/ZnFe$_2$O$_4$ 中空纳米块的（a）XRD、（b）FESEM、（c）TEM 和（d）HRTEM 图

由于 ZnO 作为两性氧化物可以在 NaOH 溶液中溶解，因此可以通过用 NaOH 溶液处理 h-Ag/ZnO/ZnFe$_2$O$_4$ 的方法获得 Ag/ZnFe$_2$O$_4$ 复合物。Ag/ZnFe$_2$O$_4$ 复合物的结构和形貌表征如图 5.9 所示。图 5.9（a）为其 XRD 图谱，表明该结构含有 Ag（JCPDS No. 65-2811）和 ZnFe$_2$O$_4$（JCPDS No. 22-1012）。其 SEM 图（见图 5.9（b））表明 Ag/ZnFe$_2$O$_4$ 复合物在去除了 ZnO 后保持了块状结构。其 BET 吸附-脱附分析和 BJH 孔径分布结果如图 5.10 所示。其 BET 比表面积为 95.23 m$^2\cdot$g^{-1}，与未处理的 h-Ag/ZnO/ZnFe$_2$O$_4$ 相比其孔径分布更宽。

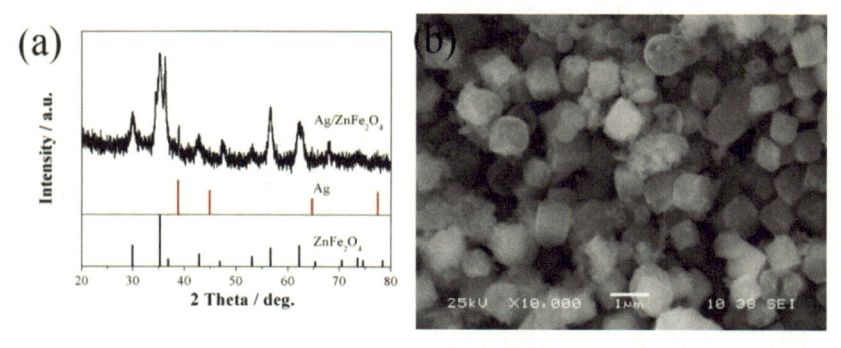

图 5.9 Ag/ZnFe$_2$O$_4$ 的（a）XRD 和（b）SEM 图

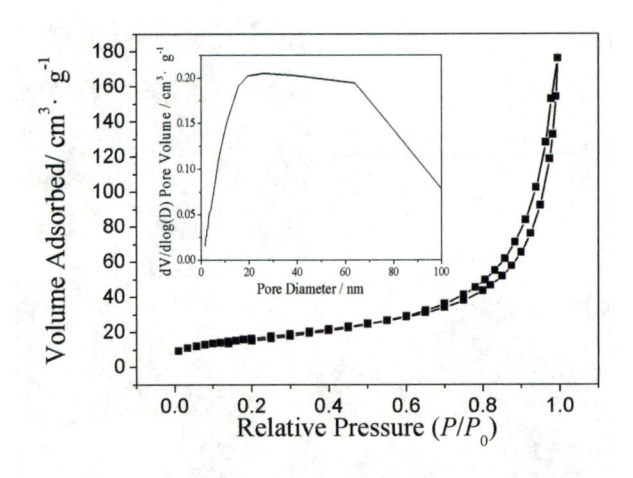

图 5.10　Ag/ZnFe$_2$O$_4$ 的 N$_2$ 吸附-脱附分析，插图为其 BJH 孔径分布图

5.3.2 光催化性能

样品的光吸收性能用紫外-可见漫反射光谱测试。如图 5.11 所示，二元复合物 ZnO/ZnFe$_2$O$_4$ 在约 700 nm 处具有明显的可见光吸收，这一结果与文献报道的 ZnO 掺杂的 ZnFe$_2$O$_4$ 样品一致[263]。Ag/ZnO/ZnFe$_2$O$_4$ 表现出更强的可见光吸收性能，这是由于银纳米颗粒的表面等离子共振性能[283]，并且中空结构的 h-Ag/ZnO/ZnFe$_2$O$_4$ 表现出更强的可见光吸收能力。

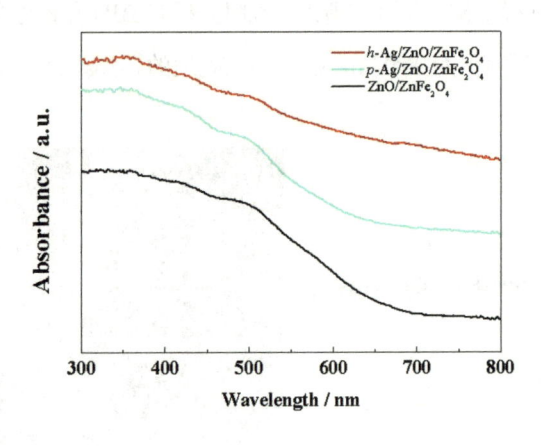

图 5.11　h-Ag/ZnO/ZnFe$_2$O$_4$、p-Ag/ZnO/ZnFe$_2$O$_4$ 和 ZnO/ZnFe$_2$O$_4$ 的紫外-可见漫反射图谱

第5章 多孔/中空 Ag/ZnO/ZnFe$_2$O$_4$ 三元复合物的制备及其光催化性能研究

在模拟太阳光下用 MB 溶液作为模拟污染物，测试了 ZnFe$_2$O$_4$ 基光催化剂的性能。结果如图 5.12（a）所示，其光催化性能按照 ZnO/ZnFe$_2$O$_4$、Ag/ZnFe$_2$O$_4$、p-Ag/ZnO/ZnFe$_2$O$_4$ 和 h-Ag/ZnO/ZnFe$_2$O$_4$ 的顺序依次增强。以 ZnO/ZnFe$_2$O$_4$ 作为光催化剂，在光照 100 min 后只有 20% 的 MB 被降解，说明其光催化性能不高。Ag/ZnFe$_2$O$_4$ 的光催化性能有所提高，同样的反应时间 MB 的降解率达到了 43%。与上述二元复合物相比，三元复合物 p-Ag/ZnO/ZnFe$_2$O$_4$ 和 h-Ag/ZnO/ZnFe$_2$O$_4$ 的光催化性能明显提高，分别达到了 60% 和 93%。假设染料光催化降解过程服从一级动力学方程（$c = c_0 e^{-kt}$，k 是速率常数），那么可以用 $\ln(c/c_0)$-t 作一条线，如图 5.12（b）所示。这条线的拟合的数据 k（min^{-1}）和方差 R^2 列在表 5.1 中，结果表明该降解反应符合一级动力学。中空的 h-Ag/ZnO/ZnFe$_2$O$_4$ 在所有的催化剂中具有最好的光催化效果。h-Ag/ZnO/ZnFe$_2$O$_4$ 的 k 值（0.021 min^{-1}）是 p-Ag/ZnO/ZnFe$_2$O$_4$（0.007 6 min^{-1}）的 2.8 倍、Ag/ZnFe$_2$O$_4$（0.003 3 min^{-1}）的 6.4 倍和 ZnO/ZnFe$_2$O$_4$（0.001 7 min^{-1}）的 12.4 倍。这说明中空结构催化剂由于不同组分的协同作用减少了光生电子的复合概率，并且较大的比表面积（63.51 m^2·g^{-1}）也为光催化反应提供了更多的活性位点。

为了考察银的掺杂量对于 Ag/ZnO/ZnFe$_2$O$_4$ 复合物光催化性能的影响，实验中用于掺杂的 AgNO$_3$ 溶液用量分别为 0.75 mL、3.0 mL 和 4.0 mL。相应的样品分别命名为 h-Ag/ZnO/ZnFe$_2$O$_4$-1、h-Ag/ZnO/ZnFe$_2$O$_4$-2 和 h-Ag/ZnO/ZnFe$_2$O$_4$-3。样品的 EDX 光谱（见图 5.13）表明，h-Ag/ZnO/ZnFe$_2$O$_4$-1、h-Ag/ZnO/ZnFe$_2$O$_4$-2 和 h-Ag/ZnO/ZnFe$_2$O$_4$-3 的银含量的质量分数分别为 1.15%、2.70% 和 3.94%。银含量不同的催化剂，其性能如图 5.12（c）和 5.12（d）所示。首先，随着银含量的增加，光催化性能增加，随着银含量继续增加，光催化性能反而降低。降解过程符合一级动力学过程，k 值按照 h-Ag/ZnO/ZnFe$_2$O$_4$-3、h-Ag/ZnO/ZnFe$_2$O$_4$-1、h-Ag/ZnO/ZnFe$_2$O$_4$-2 和 h-Ag/ZnO/ZnFe$_2$O$_4$ 的顺序增加（见表 5.1）。这一结果表明，银含量对于 Ag/ZnO/ZnFe$_2$O$_4$

复合物的光催化性能有重要影响，并且在质量分数为 2.17%时表现出最佳的光催化性能。银的掺杂对光催化性能的影响的结果与 Ag-TiO$_2$ 一致，这说明银纳米颗粒的功函对于催化剂性能有重要影响[284]。

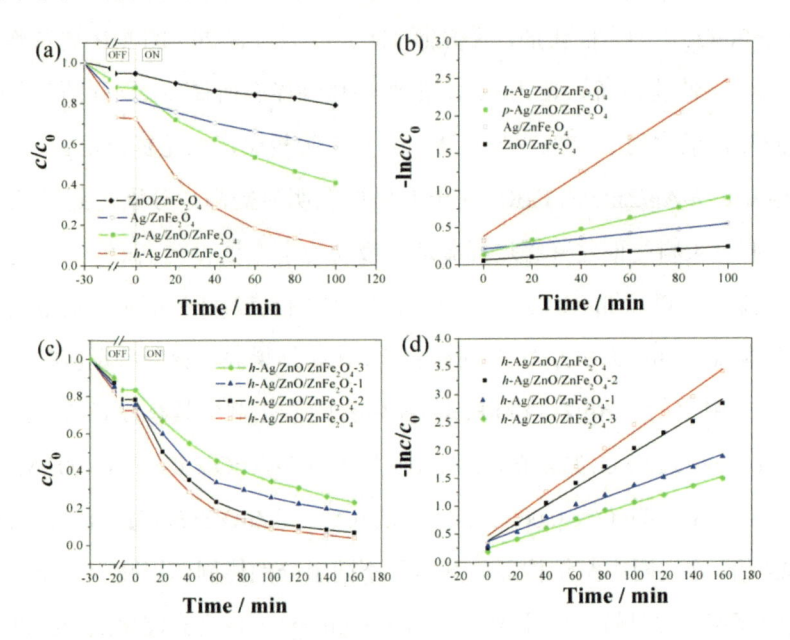

图 5.12 不同形貌催化剂对 MB 的（a）降解曲线、（b）准一级动力学曲线；银掺杂浓度对 Ag/ZnO/ZnFe$_2$O$_4$ 光催化性能的影响，（c）降解曲线，（d）准一级动力学曲线

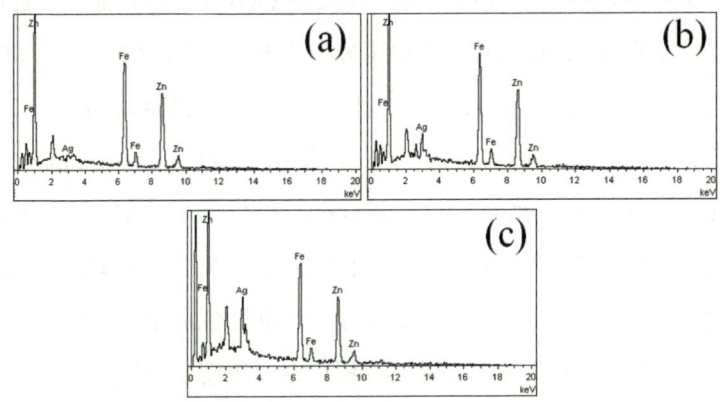

图 5.13 （a）h-Ag/ZnO/ZnFe$_2$O$_4$-1、（b）h-Ag/ZnO/ZnFe$_2$O$_4$-2、（c）h-Ag/ZnO/ZnFe$_2$O$_4$-3 的 EDX 谱图

表 5.1　不同催化剂降解 MB 的准一级动力学参数

样品	$K/(\text{min}^{-1})$	R^2
h-Ag/ZnO/ZnFe$_2$O$_4$	0.0210	0.996
p-Ag/ZnO/ZnFe$_2$O$_4$	0.0076	0.994
Ag/ZnFe$_2$O$_4$	0.0033	0.997
ZnO/ZnFe$_2$O$_4$	0.0017	0.971
h-Ag/ZnO/ZnFe$_2$O$_4$-1	0.0097	0.990
h-Ag/ZnO/ZnFe$_2$O$_4$-2	0.0158	0.990
h-Ag/ZnO/ZnFe$_2$O$_4$-3	0.0080	0.992

TOC 分析证明了光催化降解 MB 过程中的矿化程度[285]。如图 5.14 所示，在模拟太阳光照射 160 min 后 h-Ag/ZnO/ZnFe$_2$O$_4$ 的 TOC 去除率达到了 87%，这表明在 MB 的降解过程中发生了矿化，并且 MB 在 Ag/ZnO/ZnFe$_2$O$_4$ 催化剂作用下可以完全降解。

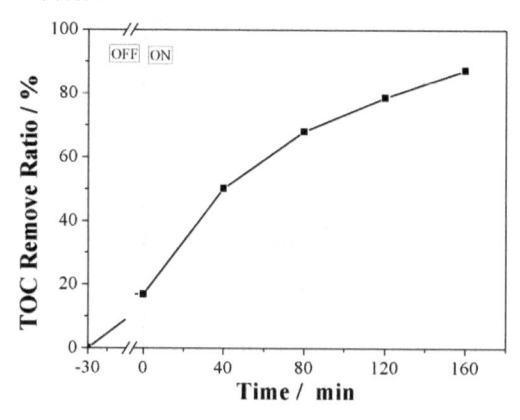

图 5.14　h-Ag/ZnO/ZnFe$_2$O$_4$ 的 TOC 降解曲线

对 h-Ag/ZnO/ZnFe$_2$O$_4$ 催化剂降解 MB 过程进行了重复实验，用于评价催化剂的稳定性。结果如图 5.15（a）所示，经过 5 次降解循环催化剂的活性略有降低。这说明催化剂具有很好的稳定性。图5.15（b）给出了 h-Ag/ZnO/ZnFe$_2$O$_4$ 的磁滞回线，说明该材料具有超顺磁性。如图5.15（b）中的插图所示，当磁铁靠近反应溶液时，h-Ag/ZnO/ZnFe$_2$O$_4$ 催化剂可以快速地吸到磁铁一边，同时溶液变得澄清。这一结果表明，Ag/ZnO/ZnFe$_2$O$_4$ 催化剂可以很容易通过磁铁分离。

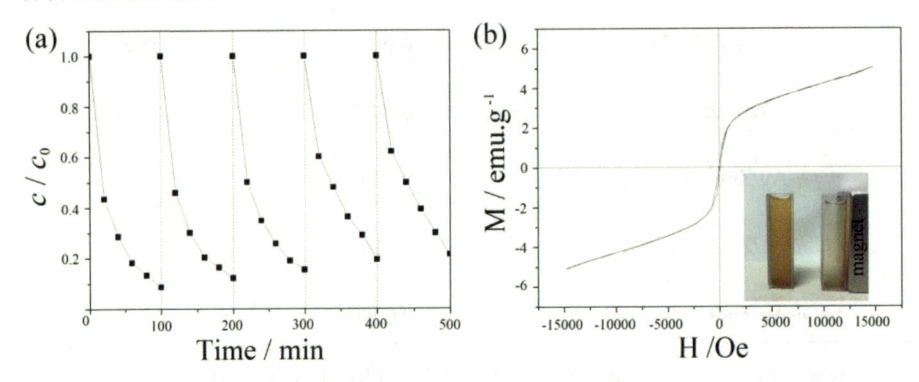

图5.15 （a） h-Ag/ZnO/ZnFe$_2$O$_4$ 的循环曲线和（b）磁滞回线，插图为催化剂分离效果的数码照片

通常认为光催化过程中空穴（h$^+$）、羟基自由基（·OH）和超氧自由基（O$_2^-$）对有机污染物的降解起重要作用[155]。通过在光催化反应体系中添加少量的EDTANa$_2$（空穴俘获剂）、叔丁醇（t-BuOH，·OH俘获剂）和对苯醌（BZQ，O$_2^-$俘获剂），可确定主要的催化活性物质并了解光催化机理[156-157]。如图5.16所示，在 t-BuOH、BZQ 和 EDTANa$_2$ 添加后 h-Ag/ZnO/ZnFe$_2$O$_4$ 的催化反应速率常数分别为 0.020 9 min^{-1}、0.012 7 min^{-1} 和 0.006 6 min^{-1}。从中可看出，BZQ 和 EDTANa$_2$ 的添加明显降低了反应速率，而 t-BuOH 对光催化速率基本没有影响。这一结果表明光催化过程中 h$^+$ 和 O$_2^-$ 起到主要作用，而·OH 不是该催化剂产生的活性物质。

示意图5.17显示了催化剂在光催化过程中的物质转移过程。作为窄带隙的半导体，ZnFe$_2$O$_4$ 可以吸收可见光[252]，但是由于光生电子和空穴的快速复合，ZnFe$_2$O$_4$ 的光催化效果较差[286]。当 ZnFe$_2$O$_4$ 与 ZnO 形成复合物时，

第 5 章　多孔/中空 Ag/ZnO/ZnFe$_2$O$_4$ 三元复合物的制备及其光催化性能研究

由于 ZnFe$_2$O$_4$ 的价带和导带与 ZnO 相比更正，可以形成 II 型的能带匹配。因此 ZnO/ZnFe$_2$O$_4$ 形成的二元复合物可以更有效地分离光生电子和空穴，使电子转移到 ZnO 的导带，而空穴留在 ZnFe$_2$O$_4$ 的价带。

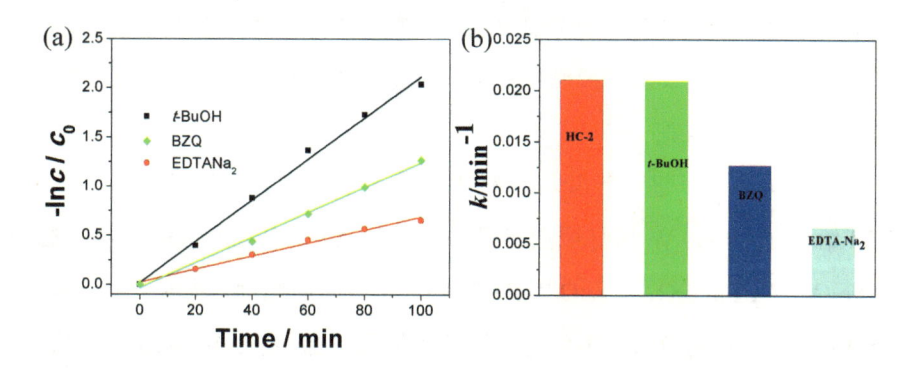

图 5.16　添加俘获剂后 h-Ag/ZnO/ZnFe$_2$O$_4$ 的（a）准一级动力学曲线和（b）速率常数图

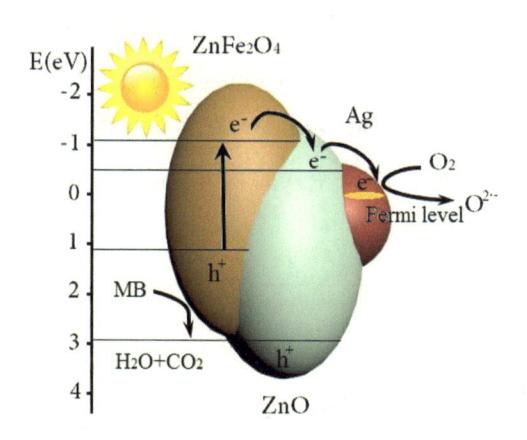

图 5.17　Ag/ZnO/ZnFe$_2$O$_4$ 对有机污染物的光催化降解机理

当 ZnO 经金属银修饰后，部分光生电子被银和 ZnO 之间的肖特基势垒俘获，从 ZnO 的价带转移到银纳米颗粒的表面[287]。ZnO 的功函约 5.2 eV，第一电子亲和能约 4.3 eV，而银的功函约 4.26 eV[288-289]。由于 ZnO 的功函较大，其费米能级比银低。这导致了电子从银向 ZnO 转移，直到两个系统达到平衡，并形成新的费米能级（E_f）[290-291]。电子的转移继续进行直到银修饰的 ZnO 的费米能级移向更负的水平，并最终与银颗粒达到平衡[292-294]。

由于 ZnO 导带的能级水平比新形成的费米能级高，光生电子可以转移到金属银纳米颗粒。接着银俘获的电子可以被吸附的 O_2 俘获形成 O_2^-。因此光生电子和空穴复合概率进一步降低，导致 Ag、ZnO 和 $ZnFe_2O_4$ 形成复合物增强了光催化活性。根据一些课题组报道，如果添加的金属过多，这些金属颗粒会成为光生电子和空穴复合的中心，从而降低光催化活性[295-298]。这与我们报道的结果一致。

光电响应可以进一步验证材料中光生电子和空穴分离的效率[299]。图 5.18 给出了在 500 W 氙灯照射下四种样品 p-Ag/ZnO/$ZnFe_2O_4$、h-Ag/ZnO/$ZnFe_2O_4$、Ag/$ZnFe_2O_4$ 和 ZnO/$ZnFe_2O_4$ 的光电流响应图谱。从图中可以看出，ZnO/$ZnFe_2O_4$ 和 Ag/$ZnFe_2O_4$ 的光电流信号较弱，三元复合物 Ag/ZnO/$ZnFe_2O_4$ 的光电流信号均比二元复合物强。这一结果表明，三元复合物 Ag/ZnO/$ZnFe_2O_4$ 的光生电子-空穴的有效分离、表面复合概率明显降低。值得注意的是，中空结构的 h-Ag/ZnO/$ZnFe_2O_4$ 比多孔结构的 p-Ag/ZnO/$ZnFe_2O_4$ 具有更高的光流响应。并且 h-Ag/ZnO/$ZnFe_2O_4$ 在本章报道的复合物中具有最强的光电流信号，表明其光生电子-空穴的利用率最高。这一结果与我们在光催化实验中获得的结果一致。

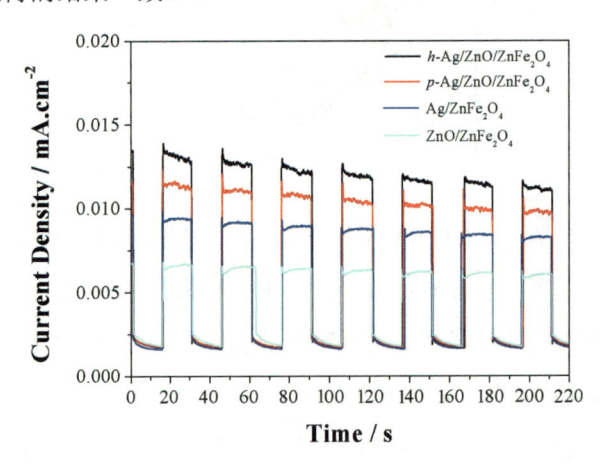

图 5.18 h-Ag/ZnO/$ZnFe_2O_4$、p-Ag/ZnO/ZFe_2O_4、Ag/$ZnFe_2O_4$ 和 ZnO/$ZnFe_2O_4$ 的光电流响应图

5.4 本章小结

通过一种可控煅烧 Ag-ZnPBA 前驱体的方法获得了三元复合物 Ag/ZnO/ZnFe$_2$O$_4$ 的多孔和中空纳米块。Ag/ZnO/ZnFe$_2$O$_4$ 中空纳米块的光催化活性比多孔的 Ag/ZnO/ZnFe$_2$O$_4$、ZnO/ZnFe$_2$O$_4$ 以及 Ag/ZnFe$_2$O$_4$ 纳米块高。这种新颖的复合方式使结构中同时具备 II 型能带匹配和肖特基势垒，导致光生电子-空穴更有效地分离，进而提高了光催化性能。另外由于具有磁性，Ag/ZnO/ZnFe$_2$O$_4$ 催化剂可以通过磁铁进行回收利用，这对于实际应用具有重要意义。这一研究表明，利用 MOF 作为前驱体制备多组分、多功能的复合材料可操作性更强。该方法不仅实现了颗粒的形貌和尺寸可调，并且复合物成分也可以简单地调控。这使其在环境和能源领域的应用有巨大的潜能。

第6章 分级结构的 Fe/Fe$_3$C@N doped C 复合物的制备及其 HER 性能研究

6.1 引言

与传统的化石能源相比，氢是未来最有前途的清洁、可再生能源[300]。因此制氢技术受到了重视，HER 是目前最简单、最有吸引力的水分解制氢技术[301]。但是商用的 HER 催化剂均含有昂贵的 Pt 等贵金属材料[302-305]。在过去的十年中，研究人员开发非贵金属 HER 催化剂[306] 的过程中，发现过渡金属及其合金、硫化物、磷化物也可作为 HER 催化材料，例如：Mo 基[307-309]、Co 基复合材料[310]、WS[311]、Ni$_2$P[312]以及 FeCo 合金[313] 等。由于铁、钴和镍在地壳中丰度较高，将其设计成高效的 HER 电极材料对于 HER 技术的大规模应用具有重要意义。

通常认为形成碳基复合材料是提高纳米材料 HER 性能的有效途径[314]。Bao 的课题组通过在氮气保护下煅烧的方法，将 Fe 的纳米颗粒封装在碳纳米管内获得了性能优异且稳定的 HER 电极材料[315]。Zheng 等在氮气保护下煅烧多面体形 Zn-MOF 微块，获得了氮掺杂的多孔炭材料，并保持了多面体形貌[316]。Wu 的课题组用 MOF 和二氰胺混合煅烧得到了新颖的氮掺杂石墨烯纳米管结构。并且提出了类似于碳纳米管的原位催化生长的 CVD 生长机理[317-318]。氮掺杂的石墨烯与金属形成的复合物可以大幅提高电极材料的电催化活性[319-321]。这可能是因为氮掺杂石墨烯提供了更多的活性位点和结构缺陷，并且保护催化剂使其可以在各种介质中进行[318, 322-323]。上述研究表明，简单的 MOF 材料煅烧就可制备高效、稳定的 HER 催化材料。

本章通过在氮气下煅烧多面体形 Zn$_3$[Fe(CN)$_6$]$_2$·xH$_2$O 微块制备了具有分

级结构的 Fe/Fe₃C@N doped C 复合材料，进而测试了其 HER 性能。

6.2 实验部分

6.2.1 材料

本实验用到的所有化学试剂均为分析纯，购自国药集团。实验中用到的水均为去离子水。

6.2.2 分级核壳结构 Fe/Fe₃C@N doped C 纳米复合物的制备

按照第 4 章的实验方法制备多面体形 $Zn_3[Fe(CN)_6]_2·xH_2O$ 前驱体。

将 100 mg 前驱体颗粒置于瓷舟，放入管式炉中，在氮气氛围下以 5 °C·min^{-1} 的升温速率加热到 800 °C，恒温 1 h，之后停止加热自然降温到室温。样品记为 Fe/Fe₃C@N doped C-1。按照上述过程温度升高到 800 °C 后立即降温，获得的样品记为 Fe/Fe₃C@N doped C-0。保温 2 h 的样品记为 Fe/Fe₃C@N doped C-2。

6.2.3 仪器

物相结构由 X 射线衍射（XRD，Rigaku D/MAX 2400，Cu-$K_α$ $λ$ = 1.540 6 Å，扫描速率 2 °/min）表征。样品的形貌、粒径和微观结构用 FESEM（Hitachi S-4800，加速电压 20 kV）和 TEM（JEM-2100，加速电压 200 kV）表征。成分分析通过 SEM 附带的能谱仪测试（EDX）进行表征。元素分析（EA）在 Perkin-Elmer 240C 上测试。金属含量用等离子发射光谱（ICP-OES，Vista-MPX）分析。元素 Mapping 在 Tecnai G2 F30 S-Twin TEM 上测试。拉曼光谱测试通过 DXR 拉曼光谱仪用 532 nm 激光源测试。X 射线光电子能谱（XPS）在 Thermo ESCALAB 250XI 上测定。热重-差热（TG-DSC）分析在氮气保护下测试，升温速率 22.5 °C/min（NETZSCH5 STA 449F3，Germany）。

6.2.4 电化学性能测试

HER 性能由三电极系统的 CHI 760D 电化学工作站（上海辰华）在室温下进行测试。分别用饱和甘汞电极和 Pt 电极作为参比电极和对电极。将样品覆盖在玻碳电极（GCE 直径 3 mm）上作为工作电极。在覆盖样品之前，玻碳电极用 0.05 μm 的 Al_2O_3 粉末抛光，抛光后在乙醇中超声清洗，并在室温下干燥。在涂覆之前，取 5.0 mg 的催化剂粉末加入 0.98 mL 的无水乙醇中，同时加入 20 μL 的萘酚溶液，之后混合溶液经超声处理 30 min。取上述溶液 5 μL 滴到玻碳电极上，室温下自然干燥。电化学测试前，电解液（0.5 M H_2SO_4）用氮气处理 60 min。线性扫描伏安法的扫描速率为 2 mV·s^{-1}。

6.3 结果与讨论

6.3.1 结构和形貌表征

多面体形 $Zn_3[Fe(CN)_6]_2·xH_2O$ 微-纳米块的结构和形貌在前面章节已经表征，由 6 个正方形和 8 个六边形构成。$Zn_3[Fe(CN)_6]_2·xH_2O$ 前驱体的 TG-DSC 分析结果如图 6.1 所示，前驱体在 100 °C 和 600 °C 分别有一次失重，分别对应结晶水和 CN 的分解。由于 Zn 在惰性气氛下高温蒸发的特性，600 °C 之后的持续失重通常认为是 Zn 的蒸发[324]。

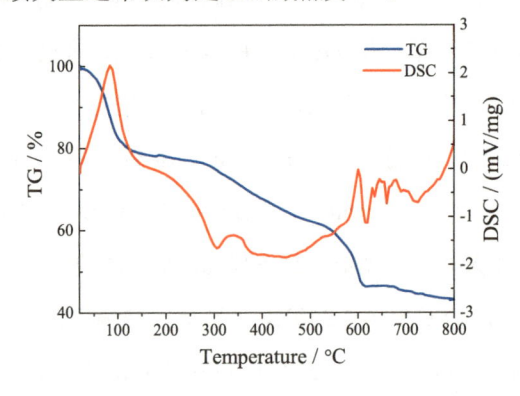

图 6.1 前驱体 $Zn_3[Fe(CN)_6]_2·xH_2O$ 在氮气中煅烧的 TG-DSC 图

第 6 章 分级结构的 Fe/Fe₃C@N doped C 复合物的制备及其 HER 性能研究

将前驱体在氮气氛围中 800 ℃ 煅烧 1 h，获得黑色粉末。图 6.2 为样品的 XRD 图，图中在 24.6° 附近有一个宽的衍射峰，这对应于石墨晶格的（002）面[325]。谱图中位于 $2\theta = 37.8°$，40.5°，42.8°，43.7°，44.6°，45.0°，49.1° 和 51.8° 处的衍射峰可归属于 Fe₃C（JCPDS，No. 89–2867）。谱图中位于 $2\theta = 44.7°$ 和 65.0° 处的衍射峰可归属于 α-Fe（JCPDS，No. 87–0722）。结果表明，该样品为石墨、α-Fe 和 Fe₃C 的复合物。

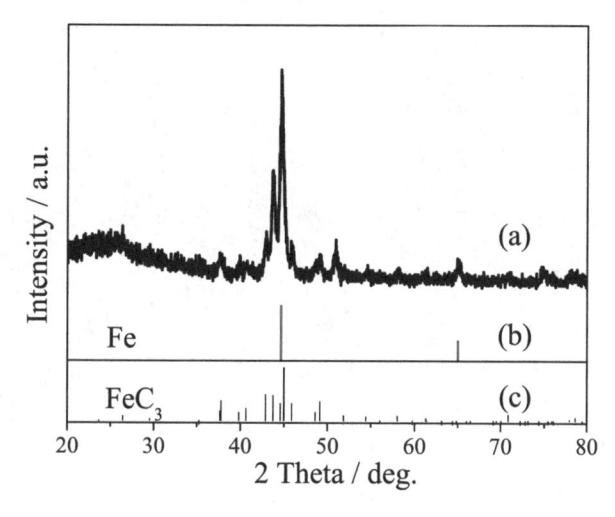

图 6.2 样品 Fe/Fe₃C@N doped C-1 的 XRD 谱图

煅烧处理后样品 Fe/Fe₃C@N doped C-1 的形貌如图 6.3 所示。图 6.3（a）是样品的 FESEM 图片，图中可以看出产物保持了均匀的多面体形貌，粒径约为 1 μm。在更高倍数下观察（见图 6.3（b）），可以发现多面体微块的表面有很多棒状突起。样品的 TEM 图片（见图 6.3（c））提供了更多的形貌细节。从图中可以看出，多面体微块由许多粒径在 10～50 nm 的纳米颗粒堆积而成，纳米颗粒之间存在明显的多孔结构。在更高倍数下（见图 6.3（d））可以明显地看到构成多面体微块的纳米颗粒具有核壳结构。多面体微块的表面有一些向外生长的碳纳米管，并且纳米管的头部有一个金属颗粒。这一结构类似于在基底上 CVD 法原位生长的碳纳米管。核壳结构纳米颗粒的 HRTEM 图（见图 6.3（d）中的插图）表明外壳的间距为 0.34 nm，与碳的

（002）晶面相符，而内核的晶格条纹间距为 0.21 nm，与 Fe₃C 的（211）晶面相符。由于 α-Fe 的（311）晶面与 Fe₃C 的（211）晶面非常接近，很难从 HRTEM 上进行区分。[326-327] Fe、C、N 元素的 Mapping 图可明显看出三元素在颗粒中均匀分布，碳管的头部有铁元素聚集。结合 EDX 分析结果，可以确定通过简单的加热技术成功地获得了具有分级结构的氮掺杂 C、α-Fe、Fe₃C 复合物。

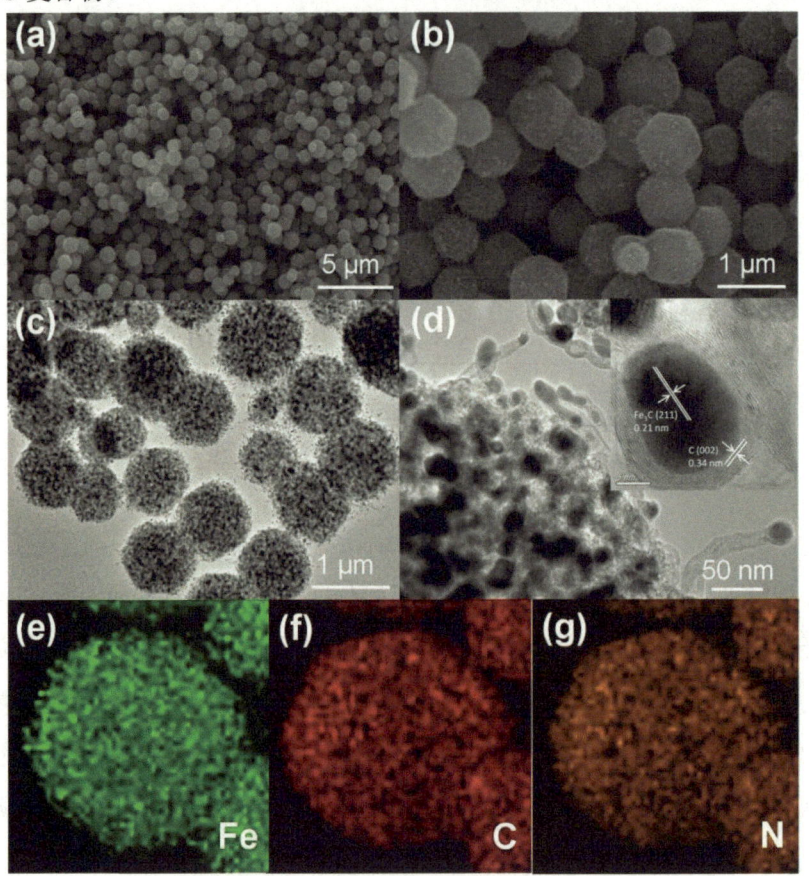

图 6.3 （a，b）样品的 FESEM 和（c，d）TEM 图，（d）中的插图为 HRTEM 图，（e～g）分别为 Fe、C、N 的元素 Mapping 图

样品的组分分别用 EDX 和 EA 进行表征，结果表明样品的成分为 Fe、C 和 N，其中铁含量为 71.17%，C 和 N 的质量分数分别为 27.12%和 1.61%。

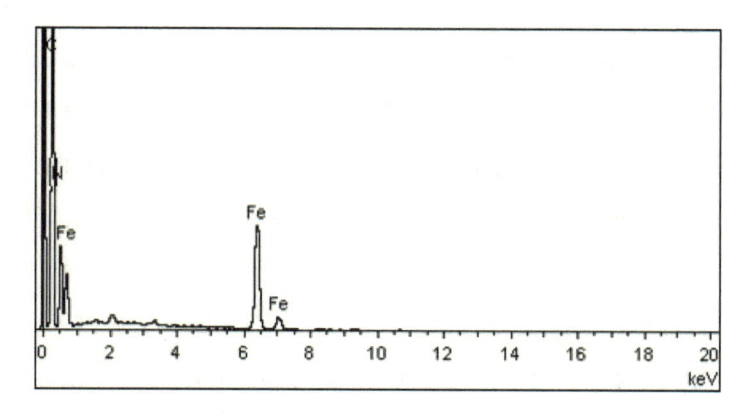

图 6.4 样品 Fe/Fe₃C@N doped C-1 的 EDX 图谱

XPS 谱图（图 6.5）表明样品 Fe/Fe₃C@N doped C-1 中含有 Fe、C、N 三种元素，其中铁元素的氧化数分别为零和Ⅲ，C 存在 C 1s 和 C-N 两种状态，氮元素分为三种状态，分别为吡啶氮（398.4 eV）、吡咯氮（400 eV）和石墨氮（401.1 eV）[328]。氮掺杂可为 HER 催化反应提供更多的活性位点。

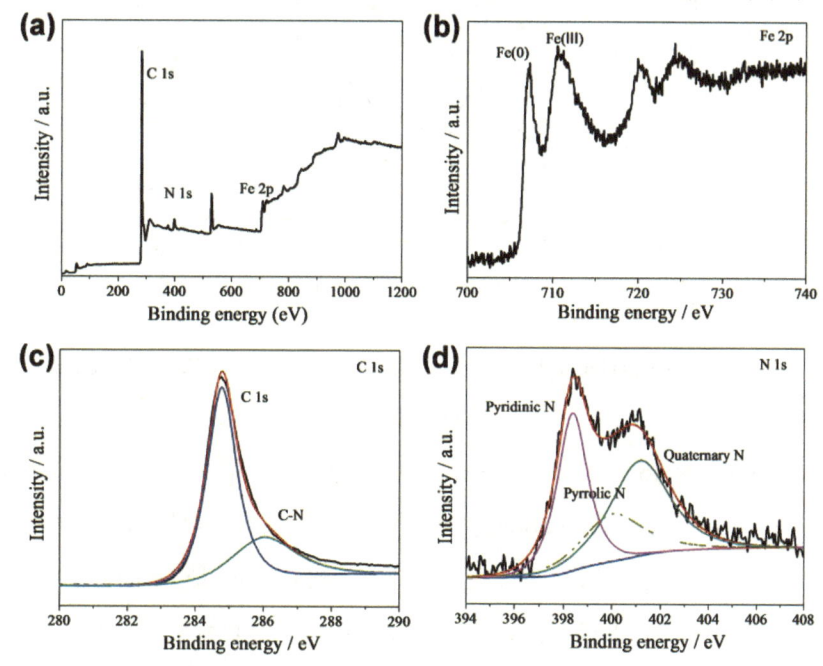

图 6.5 （a）样品 Fe/Fe₃C@N doped C-1 的 XPS 图谱：（a）总谱、（b）Fe 2p、（c）C 1s、（d）N 1s

拉曼光谱广泛用于表征碳材料的结构特征，如缺陷和无序结构。Fe/Fe3C@N doped C-0、Fe/Fe3C@N doped C-1、Fe/Fe3C@N doped C-2 样品的拉曼图谱（见图 6.6）显示，所有的样品的拉曼图谱中均含有两个突出的峰，D 峰 1 350 cm^{-1} 和 G 峰 1 580 cm^{-1}。[329] D 峰与结构中的缺陷和石墨结构有序性的破坏相关，而 G 峰通常与单晶石墨烯晶面中 sp^2 碳原子的 E_{2g} 声子伸缩振动相关[329-333]。随着保温时间的增加，G 峰的相对强度增高，同时 D 峰的相对强度减弱，说明保温时间增加，样品的石墨化程度增大，缺陷变少，结晶度更好。

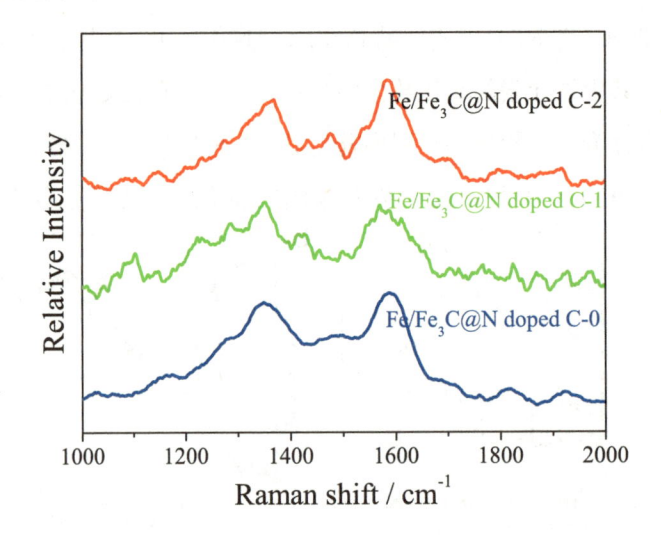

图 6.6 样品 Fe/Fe3C@N doped C-0、Fe/Fe3C@N doped C-1、Fe/Fe3C@N doped C-2 的拉曼谱图

Fe/Fe3C@N doped C-0 和 Fe/Fe3C@N doped C-2 的形貌如图 6.7 所示。图 6.7（a）是 Fe/Fe3C@N doped C-0（保温 0 h）的 SEM 照片，图中可以看出多面体颗粒比较完整，表面未见碳纳米管。图 6.7（b）是 Fe/Fe3C@N doped C-2（保温 2 h）的 SEM 照片，图中可看出颗粒有些已经破碎，表面可见一些管状物凸起。结合 Fe/Fe3C@N doped C-1 的 SEM 图（图 6.3）可以推断只有在合适的保温时间下才可在颗粒表面生成碳纳米管。

第 6 章　分级结构的 Fe/Fe₃C@N doped C 复合物的制备及其 HER 性能研究

图 6.7　在 800 °C 下保温 0 h 和 2 h 后获得样品的 SEM 图：(a) Fe/Fe$_3$C@N doped C-0，(b) Fe/Fe$_3$C@N doped C-2

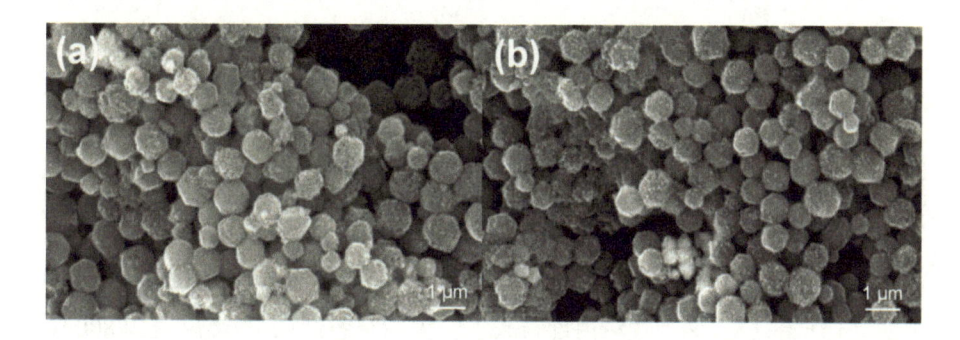

Fe/Fe$_3$C@N doped C-0、Fe/Fe$_3$C@N doped C-1、Fe/Fe$_3$C@N doped C-2 样品的磁性结果如图 6.8 所示。如图所示，三个样品的磁化强度 M_s 依次为 36.77 emu·g^{-1}、67.01 emu·g^{-1} 和 65.68 emu·g^{-1}，矫顽力 H_c 依次为 12.85 Oe、21.74 Oe、18.68 Oe。这一结果说明样品的磁性能较好，而 Fe/Fe$_3$C@N doped C-1、Fe/Fe$_3$C@N doped C-2 的饱和磁化强度比 Fe/Fe$_3$C@N doped C-0 略高。这一结果说明适当增加保温时间会提高材料的磁性能。

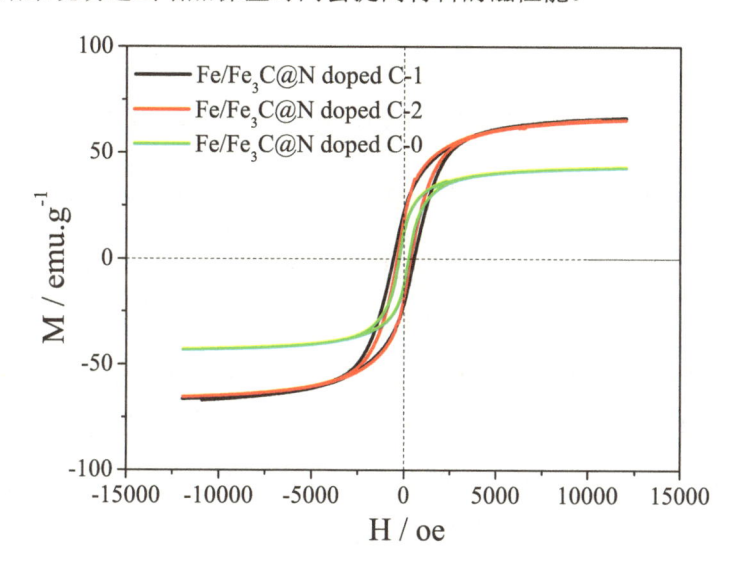

图 6.8　Fe/Fe$_3$C@N doped C-0、Fe/Fe$_3$C@N doped C-1、FeFe$_3$C@N doped C-2 的室温磁滞回线

87

6.3.2 分级结构的 Fe/Fe₃C@N doped C 复合物的形成机理

如图 6.9 所示，前驱体在氮气保护下加热时，Fe—C—N—Zn 键断裂，Fe 原子和 Zn 原子聚集并形成纳米合金颗粒，而 CN⁻ 则分解形成氮掺杂的碳和 CN 气体逸出。这导致了产品中的 C、N 含量比前驱体中低[316]。当温度升高到 600 °C 时，前驱体分解完全（见图 6.1）。随着温度继续升高，样品中的 Zn 开始蒸发，这为颗粒内部带来了较大的空腔结构。当温度达到 800 °C 时，颗粒内部形成了 C 包裹 Fe 纳米颗粒的核壳结构。在此温度下保温，多面体颗粒表面的氮掺杂碳在 Fe 的作用下向外生长，形成碳纳米管，但保温时间过长，颗粒的形貌会被破坏。

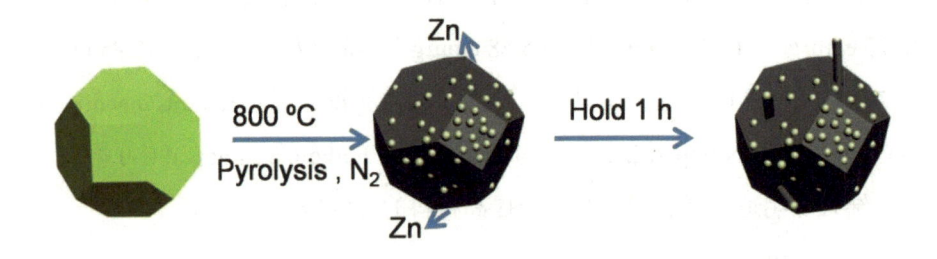

图 6.9 分级结构的 Fe/Fe₃C@N doped C 复合物生长机理

6.3.3 分级结构的 Fe/Fe₃C@N doped C 复合物的 HER 性能

由于 Fe/Fe₃C@N doped C 复合物中同时具有独特的核壳结构和碳纳米管，将其作为电极材料用于 HER 有望获得优异的性能。将 Fe/Fe₃C@N doped C-0、Fe/Fe₃C@N doped C-1 和 Fe/Fe₃C@N doped C-2 样品分别制备 HER 工作电极。HER 性能在 0.5 M 的 H_2SO_4 中测试。图 6.10（a）给出了不同时间下制备电极的 LSV 极化曲线，从图中可以看出 1 h 煅烧获得样品的过电位值最低（103 mV），在其电流密度达到 20 mA/cm² 时，过电位只需 0.26 V。其他时间下获得的样品过电位均较大，这说明 Fe/Fe₃C@N doped C-1 的 HER 性能最好。

第6章 分级结构的 Fe/Fe₃C@N doped C 复合物的制备及其 HER 性能研究

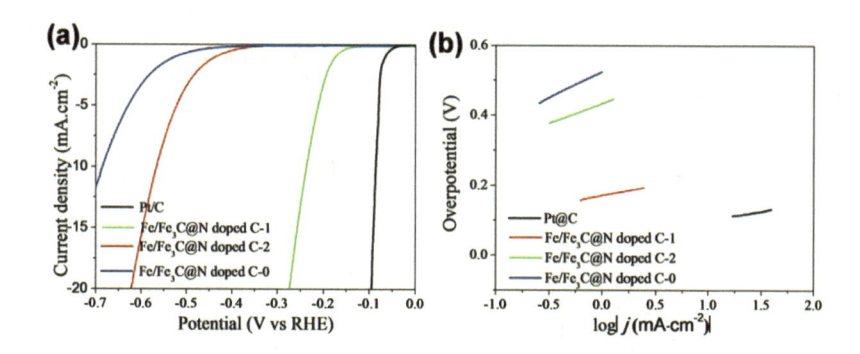

图 6.10 Fe/Fe₃C@N doped C-0、Fe/Fe₃C@N doped C-1、Fe/Fe₃C@N doped C-2、Pt/C 电极的（a）LSV 极化曲线和（b）Tafel 曲线

根据 Fe/Fe₃C@N doped C-0、Fe/Fe₃C@N doped C-1、Fe/Fe₃C@N doped C-2 电极的线性扫描伏安极化曲线，可计算并绘制样品的 Tafel 曲线。如图 6.10（b）所示，曲线的线性区域与 Tafel 公式拟合得很好（$\eta = b \log j + a$，b 是 Tafel 斜率，j 是电流密度）。Fe/Fe₃C@N doped C-1 电极的 Tafel 斜率为 59.6 mV per decade，而 Fe/Fe₃C@N doped C-0 和 Fe/Fe₃C@N doped C-2 电极的 Tafel 斜率分别为 148.3 和 113.2 mV per decade。根据催化剂的 Tafel 斜率为 59.6 mV per decade，据此推测其 HER 机理为 Volmer-Heyrovsky 机制[334-335]。Fe/Fe₃C@N doped C-1 电极的 j_0 值为 5.8×10^{-2} mA/cm²，是 Fe/Fe₃C@N doped C-0 电极 j_0 值（2.8×10^{-2} mA/cm²）的 2 倍。

在 0.5 M H₂SO₄ 溶液中对 Fe/Fe₃C@N doped C-1 进行了 1 000 次 LSV 极化测试，用于评价电催化剂的稳定性，其结果如图 6.11 所示。结果显示 1 000 个循环后，电极的催化性能有所下降。

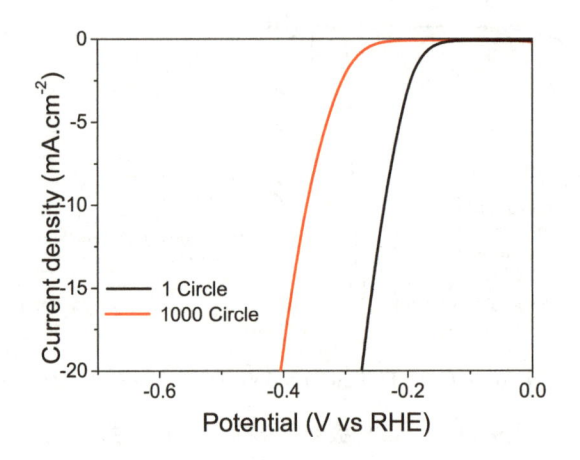

图 6.11 Fe/Fe₃C@N doped C-1 在循环第一圈和第 1000 圈后的 LSV 极化曲线

6.4 本章小结

本章以 ZnFe 普鲁士蓝类似物在氮气保护下通过简单的煅烧程序获得了具有分级结构的 Fe/Fe₃C@N doped C 复合纳米结构。该技术可使产品保持前驱体形貌，多面体微块表面形成碳纳米管结构。在多面体微块内部是由 10～50 nm 的核壳结构纳米颗粒组成，这些纳米颗粒之间存在明显的多孔结构。该材料均具有较低的过电位（103 mV）和 Tafel 斜率（59.6 mV per decade），HER 性能优异。这一工作对于研究非贵金属电化学析氢电极材料和磁存储材料具有一定的意义。

第 7 章　NiFe@N 掺杂石墨烯微管的一步法制备及其 HER 性能研究

7.1 引言

近期，镍铁基合金及其碳复合材料被用于 HER 电极材料[336-339]。其优异的性能被认为是基于火山模型的原子间协同作用降低了氢析出反应的能量[340]。第 6 章的研究结果表明，只需通过简单的煅烧过程就可将作为前驱体的 MOF 纳米材料转化为金属-碳基复合材料。利用 MOF 在形貌和成分上的优势，这一类似于原位化学气相沉积的方法可以制备多种新颖的碳基复合物[341-342]。除了保持前驱体的形貌，复合物中的碳还可在金属的催化下生长出各种新颖的形貌（如第 6 章 C 在 Fe 催化下长出了碳纳米管）[343-344]。Zheng 等利用 NiFe(CN)$_6$ 作为前驱体制备了豆荚状结构的碳纳米管（CNT），并将其用于太阳能电池的电极材料[345]。因此通过在氮气氛围中煅烧 NiFe 基 MOF 材料，有望获得形貌独特、性能优异的碳基 HER 催化材料。

本章以 Ni 硝普盐（NiFe(CN)$_5$NO·2H$_2$O）为单源前驱体，通过简单的煅烧过程，制备了氮掺杂石墨烯管包覆 NiFe 合金颗粒（NiFe-NGT）的复合 HER 催化剂。此外在不同煅烧温度下，其形貌可调节为氮掺杂石墨烯管、氮掺杂碳纳米管、NiFe@C。这一材料的结构和内在的特性使其具有较高的 HER 活性。与传统的 CVD 和 MOF 煅烧法相比，该法不外加危险的气态的碳源（如甲烷、乙烯等），通过高效地利用自身碳、氮，安全、廉价地制备了石墨烯复合材料。

7.2 实验部分

7.2.1 材料

本实验用到的所有化学试剂均为分析纯，购自国药集团。实验中用到的水均为去离子水。

7.2.2 氮掺杂石墨烯管复合物的制备

制备前驱体（NiFe(CN)$_5$NO·2H$_2$O）纳米颗粒：100 mL Ni(NO$_3$)$_2$·6H$_2$O（3 mmol）溶液在激烈搅拌下加到 100 mL 的 Na$_2$Fe(CN)$_5$NO·2H$_2$O（1 mmol）溶液中。混合溶液在室温下搅拌 1 min 后置于暗处陈化 10 h。通过高速离心分离沉淀，之后用去离子水和乙醇分别洗涤数次。沉淀在真空烘箱中 60 °C 干燥 12 h。

将前驱体转化为氮掺杂石墨烯管的复合物：前驱体在通有氮气保护的管式炉中以 5 °C·min^{-1} 的加热速率加热到 800 °C，再以 30 °C·min^{-1} 的速度降到 500 °C，然后自然冷却到室温。黑色粉末状产品用 0.5 M H$_2$SO$_4$ 处理 24 h，以除去铁镍合金。处理后用水和乙醇洗涤数次后干燥备用，并命名为 NiFe-NGT-800。如果只将温度加热到 700 °C、600 °C，其他条件相同，则得到产品分别命名为 NiFe-NGT-700 和 NiFe-NGT-600。

7.2.3 仪器

物相结构由 X 射线衍射（XRD，Rigaku D/MAX 2400，Cu-K_α λ = 1.540 6 Å，扫描速率 2 °/min）表征。样品的形貌、粒径和微观结构用 FESEM（Hitachi S-4800，加速电压 20 kV）和 TEM（JEM-2100，加速电压 200 kV）表征。成分通过 SEM 附带的能谱仪测试（EDX）测试。元素分析（EA）在 Perkin-Elmer 240C 上测试。金属含量用等离子发射光谱（ICP-OES，Vista-MPX）分析。元素 Mapping 在 Tecnai G2 F30 S-Twin TEM 上测试。

拉曼光谱测试通过 DXR 拉曼光谱仪用 532 nm 激光源测试。X 射线光电子能谱（XPS）在 Thermo ESCALAB 250XI 上测定。热重-差热（TG-DSC）分析在氮气保护下测试升温速率 22.5 °C/min （NETZSCH5 STA 449F3，Germany）。

7.2.4 电化学性能测试

HER 性能由三电极系统的 CHI 760D 电化学工作站（上海辰华）在室温下测试。分别用饱和甘汞电极和 Pt 电极作为参比电极和对电极。将样品覆盖在玻碳电极（GCE 直径 3mm）上作为工作电极。在覆盖样品之前，玻碳电极用 0.05 μm 的氧化铝粉末抛光，抛光后在乙醇中超声清洗，并在室温下干燥。在涂覆之前，取 5.0 mg 的催化剂粉末加入 0.98 mL 的无水乙醇中，同时加入 20 μL 的萘酚溶液，之后混合溶液经超声处理 30 min。取上述溶液 5 μL 滴到玻碳电极上，室温下自然干燥。电化学测试前，电解液（0.5 M H_2SO_4）用氮气处理 60 min。线性扫描伏安法的扫描速率为 2 mV·s^{-1}。

7.3 结果与讨论

7.3.1 氮掺杂石墨烯管的表征

用 Ni^{2+} 和 $[Fe(CN)_5NO]^{2-}$ 在没有表面活性剂保护的条件下可制备前驱体纳米颗粒。前驱体用 XRD、FESEM、TEM 进行表征。XRD 图（图 7.1）显示样品的所有衍射峰均与立方结构的 $NiFe(CN)_5NO·2H_2O$ （JCPDS No. 22-0772，图 7.1（a））相符。FESEM（图 7.1（b））和 TEM（图 7.1（c））结果表明其形貌为均匀的球状颗粒，直径约为 50 nm。TG-DSC 结果表明，在氮气下，该纳米颗粒在 430 °C 左右可分解完全。

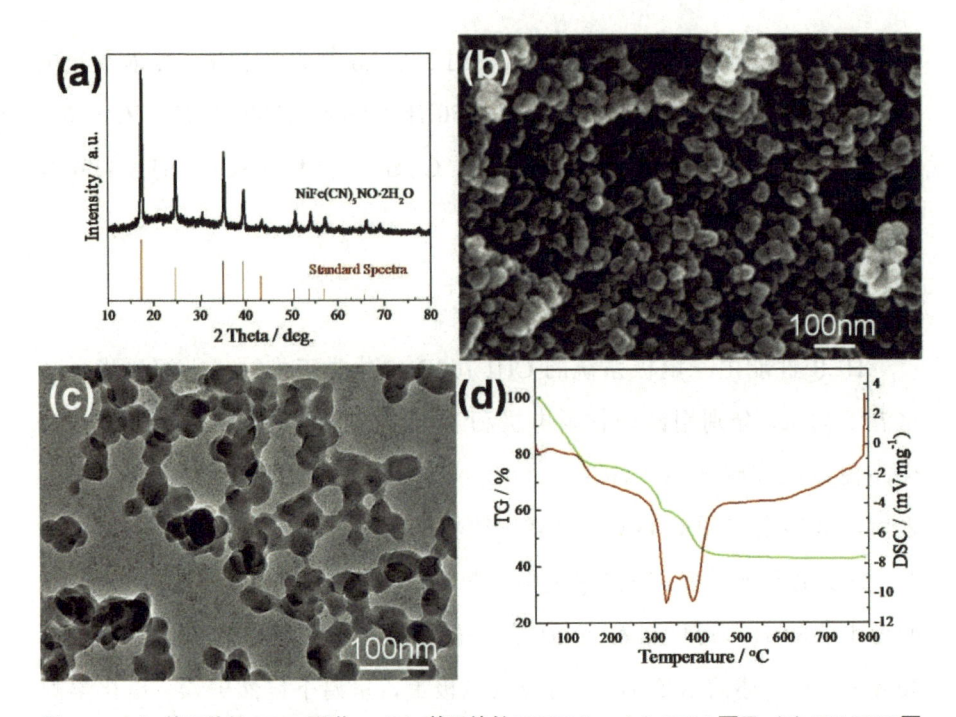

图 7.1 （a）前驱体的 XRD 图谱、（b）前驱体的 FESEM、（c）TEM 图及（d）TG-DSC 图

图 7.2 中给出了不同温度下获得样品的 XRD 图谱：NiFe-GT-800、NiFe-GT-700 和 NiFe-GT-600。NiFe-NGT-800 和 NiFe-NGT-700（见图 7.2（a）、(b)）均在 24º～26º含有一个宽峰，这被认为是石墨（002）晶格的衍射峰，说明在加热的过程中产生了石墨化的碳结构[325]。在 NiFe-NGT-800 和 NiFe-NGT-700 的图谱中还有三个尖锐的衍射峰，其 2θ分别为 43.6º、50.8º和 74.7º。这对应于铁镍合金的标准卡片 NiFe（JCPDS, No. JCPDS, No. 47-1417）。NiFe-GT-600 的 XRD 谱图中同样在 43.6º、50.8º和 74.7º处出现三个峰，而在 24º～26º处的峰（见图 7.2（c））明显较 NiFe-NGT-800 和 NiFe-NGT-700 弱。这说明该样品中的 C 结晶性较差，可能以无定形态碳存在。在三个样品的 XRD 图谱中均未出现前驱体（NiFe(CN)$_5$NO·2H$_2$O，JCPDS No. 22-0772，见图 7.1（a））的衍射峰，这与热分析的结果一致（见图 7.1（d））。

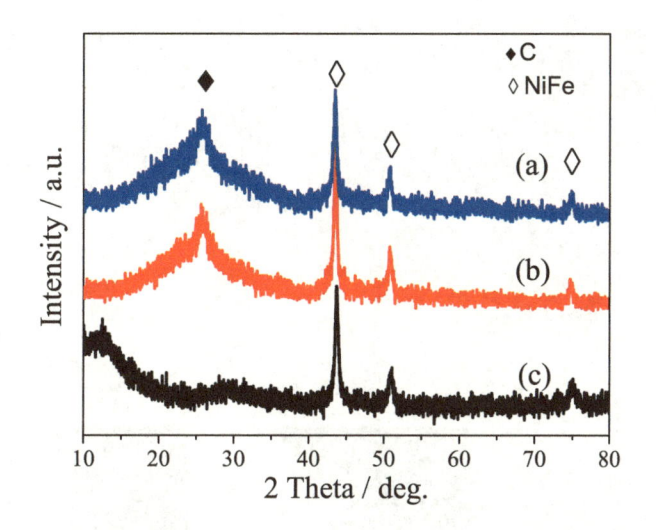

第 7 章 NiFe@N 掺杂石墨烯微管的一步法制备及其 HER 性能研究

图 7.2 （a）NiFe-GT-800，（b）NiFe-GT-700 和（c）NiFe-GT-600 的 XRD 图谱

NiFe-NGT-800 的 FESEM 图（见图 7.3（a））显示其结构中石墨烯管的直径为 1~2 μm，长度可达 10 μm。在放大的图中（见图 7.3（b））可以看到，几乎每个石墨烯管的顶端都有一个球形颗粒，更高倍数下（见图 7.3（c））可以清楚地看到球形颗粒被包裹在石墨烯中。图 7.3（d）是少数顶部没有球状颗粒的石墨烯管的高倍 FESEM 图，图中可以清楚地看到典型的石墨烯的层状形貌。NiFe-NGT-800 的形貌细节由 TEM 进一步表征。图 7.3（e）显示该结构具有典型石墨烯管状形貌，这与 FESEM 结果一致，高倍的插图可以更清楚地看到石墨烯的褶皱。从图 7.3（f）中可看到管顶部的颗粒被石墨化的碳壳包覆，而没有颗粒的一端则显示出典型的石墨烯形貌（见图 7.3（g））。在图 7.3（h）中，可以清楚地看到合金颗粒上附着的石墨烯，其 HRTEM 图（见图 7.3（i））更加明显地看出 NiFe 合金上生长着层状的石墨烯。如图 7.3（i）所示，合金颗粒表面的石墨烯层间距为 0.34 nm，与石墨的（002）晶面对应，而合金颗粒的晶格间距为 0.207 nm，与 NiFe 合金的（111）晶面对应。这一结果进一步证明了石墨烯管顶部的颗粒为被层状石墨壳包覆的 NiFe 合金，而石墨烯管就是沿着这个颗粒生长的[346]。

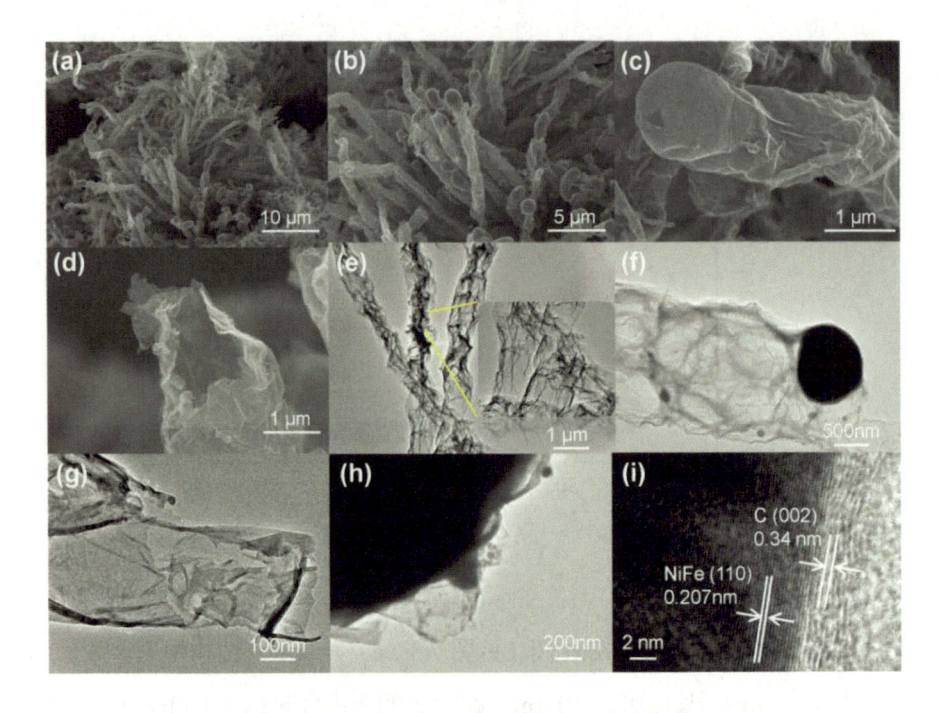

图 7.3 NiFe-NGT-800 的 FESEM 和 TEM 图

样品 NiFe-NGT-700 的 FESEM 图（见图 7.4（a）、（b））显示其形貌为竹节状的纳米管，管中含有纳米颗粒，其长度达到 50 μm。整根管子的直径是均匀的（小于 200 nm）。TEM 图片（图 7.4（c））显示了其竹节状的形貌特征，每节长度约 200 nm，并且在管子的顶端有一个被包裹的金属颗粒。其 HRTEM 图（图 7.4（d））显示碳纳米管顶部的颗粒被石墨壳包覆（002），这一点与 NiFe-NGT-800 样品相似，颗粒的晶格条纹间距为 0.207 nm 对应于 NiFe 合金的（111）晶面。我们注意到与 NiFe-NGT-800 相比，NiFe-NGT-700 的直径很小，长度也明显增加。这两个样品中，石墨烯管的直径明显比其前驱体的直径大，这说明煅烧过程中前驱体分解产生的 NiFe 合金发生了融合，温度越高，融合程度越大。石墨烯管就是由这些颗粒催化产生的，管的直径依赖于合金颗粒的直径。因此随着温度提高，石墨烯管的直径增加，由于碳、氮来源有限，管的直径越大其长度越短。在 600 ℃下煅烧的样品的 TEM 图（见图 7.4（e）、（f））中可以看到一些空心的管状结构，其直径约 50 nm，这与前驱体的直径一致。但是其 HRTEM（见图 7.4（f）中的插图）显示这些管子为无定形结构，这与其 XRD 结果一致。上述结果表明 NiFe-NGT 形貌的转化依赖于煅烧温度。

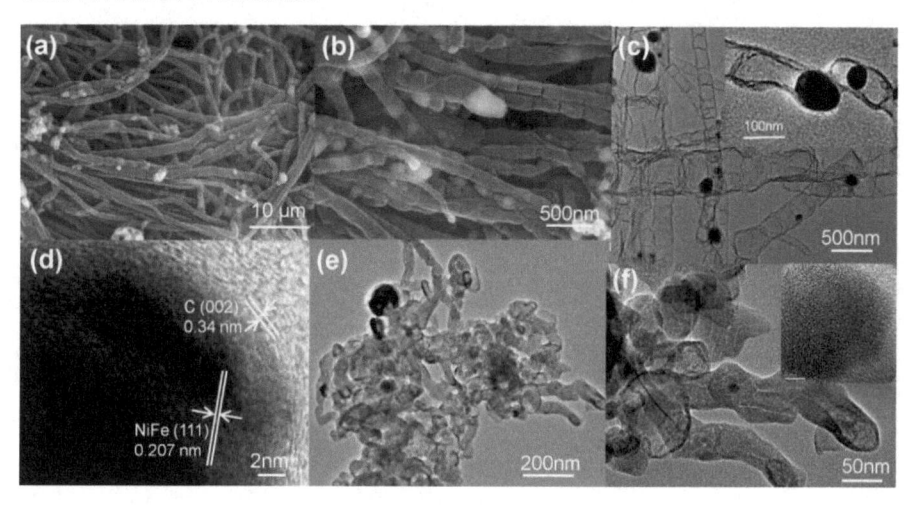

图 7.4　NiFe-NGT-700 的（a，b）FESEM、（c）TEM、（d）HRTEM 图；（e，f）样品 NiFe-NGT-600 的 TEM 图，图 f 中的插图为 NiFe-NGT-600 的 HRTEM 图

样品 NiFe-NGT-800 和 NiFe-NGT-700 的 EDX（图 7.5）、ICP 和 EA 结果表明，两个样品中 Fe:Ni 约为 1:1，这与前驱体中 Fe、Ni 比例一致（NiFe-NGT-800 中为 24.03:24.19，NiFe-NGT-700 中为 23.80:24.22）。NiFe-NGT-800 中 C 和 N 的质量分数分别为 46.96%和 4.82%，而 NiFe-NGT-700 中 C 和 N 的质量分数分别为 46.36%和 5.62%。

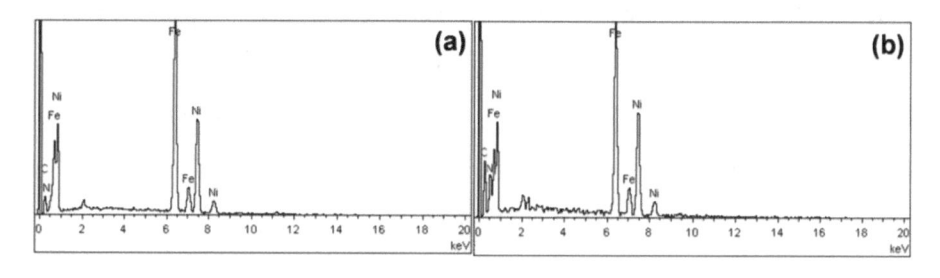

图 7.5　样品（a）NiFe-NGT-800 和（b）NiFe-NGT-700 的 EDX 谱图

Fe、Ni、C、N 四种元素在管状样品 NiFe-NGT-800 和 NiFe-NGT-700 中的分布用元素 Mapping 来表征。如图 7.6 所示，两个样品中的 Fe 和 Ni 元素均匀地分布在管子头部的金属颗粒中，而 C 和 N 在整个管中均匀分布。

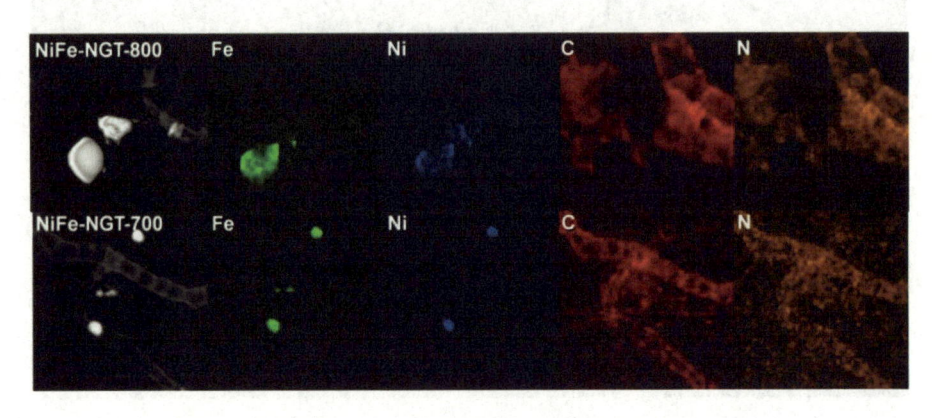

图 7.6　NiFe-NGT-800 和 NiFe-NGT-700 的元素 Mapping 图

XPS 光谱可对 NiFe-NGT-800 和 NiFe-NGT-700 样品的元素形态进行表征。图 7.7（a）和（d）分别为两样品的总谱，图中显示样品中含有 Fe、Ni、C、N 四种元素。图 7.7（b）和（e）分别是两样品的 N 1s 谱图。对其进行分峰处理后得到三个峰，分别对应于：吡啶 N（398.4 eV）、吡咯 N（399.8 eV）和石墨 N（400.7 eV）。石墨 N 处于石墨烯的碳骨架中以三个 sp^2 键与 C 原子连接，而吡啶 N 和吡咯 N 处在石墨烯边缘或缺陷处[316]。结果显示，样品中吡啶 N 和吡咯 N 在 NiFe-NGT-800 中约占 73 atm%，NiFe-NGT-700 中约占 67 atm%，占主要地位。氮掺杂也在 C 1s 谱中有所表现（见图 7.7（c）和（f）），图中四个峰分别为 sp^2 C（284.5 eV）、C—N（285.4 eV）、C=O（287.1 eV）和 O—C=O（290.5 eV）[328]。

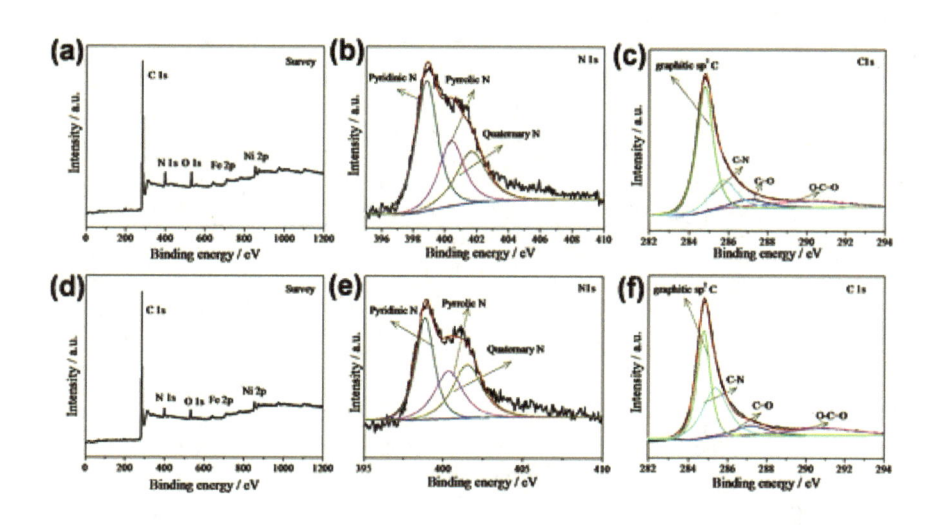

图 7.7 样品（a，b，c）NiFe-NGT-800 和（d，e，f）NiFe-NGT-700 的 XPS 图：Survey、N 1s 和 C 1s

拉曼光谱广泛用于表征碳材料的结构特征，如缺陷和无序结构。三种样品 NiFe-NGT-800、NiFe-NGT-700 和 NiFe-NGT-600 的拉曼光谱如图 7.8 所示。所有的样品的拉曼图谱中均含有两个突出的峰，D 峰 1350 cm^{-1} 和 G 峰 1580 cm^{-1} [347]。D 峰与结构中的缺陷和石墨结构有序性的破坏相关，而 G 峰通常与单晶石墨烯晶面中 sp^2 碳原子的 E_{2g} 声子伸缩振动相关[347-350]。NiFe-NGT 样品的拉曼光谱中均有较强的 D 峰和 G 峰，表明样品中碳的石墨化程度较高[351-353]，这与 HRTEM 结果一致。拉曼图谱中的 D 峰和 G 峰之间的比例（I_D/I_G）通常被认为与 sp^2 杂化区域大小相关，sp^2 杂化区域越小，I_D/I_G 比越高[329-331]。从图 7.8 可以看出，样品 NiFe-NGT-800、NiFe-NGT-700 和 NiFe-NGT-600 的 I_D/I_G 值分别约为 0.94、1.06 和 1.14。样品 NiFe-NGT-800 的 I_D/I_G 值最低，约为 0.94，这与形成大直径的石墨烯管相关，且石墨化程度较高。同时，NiFe-NGT-800 的 D 峰较另两个样品高，这说明样品中出现了更多的结构缺陷，这可能是由于高温下样品中的吡啶 N 和吡咯 N 所占比例较高[331, 333]。

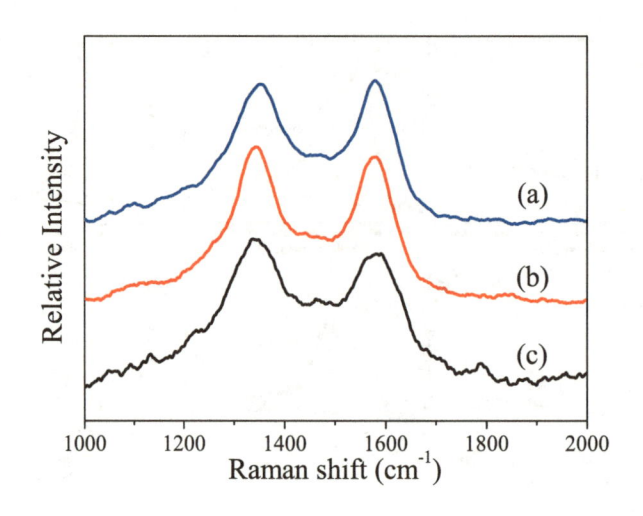

图 7.8 （a）NiFe-NGT-800、（b）NiFe-NGT-700 和（c）NiFe-NGT-600 的拉曼图

7.3.2 NiFe-NGT 复合物的形貌转化

上述结果表明，煅烧温度对产品形貌的演变起到关键作用。NiFe-NGT 复合物的生长过程如图 7.9 所示，通过分析温度对 NiFe-NGT 复合物生长过程的影响提出了可能的生长机理。煅烧过程中，前驱体在高温下分解，$Fe^{Ⅲ}$ 和 $Ni^{Ⅱ}$ 还原为 NiFe 合金颗粒[354]。而有机配体（CN^-）则分解，同时放出大量含碳氮的气体，这些气体最终会被合金颗粒俘获。在高温下，这些碳可以在 NiFe 合金颗粒的催化下通过固-液-固机制形成碳材料，如 CNTs 或石墨烯[355]。

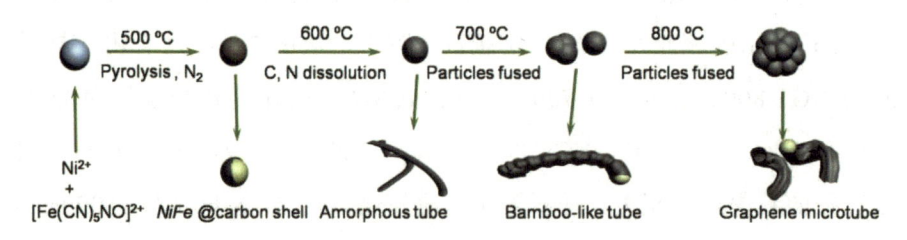

图 7.9 不同温度下石墨烯管的形成过程

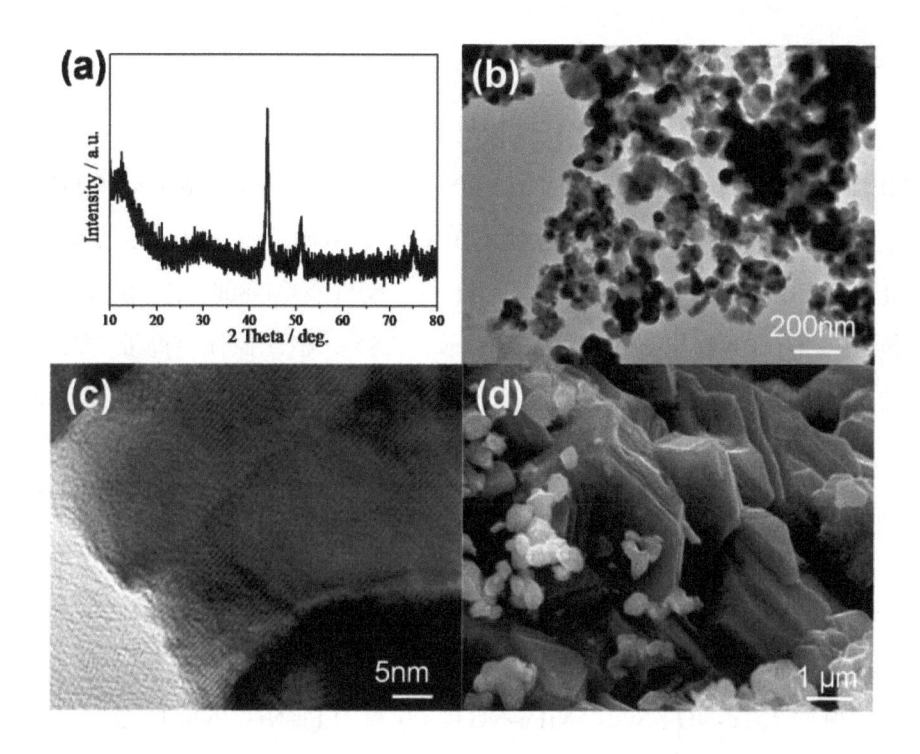

图 7.10 (a，b，c) 500 ℃下样品的 XRD、TEM 和 HRTEM 图

(d) 900 ℃下样品的 SEM 图

温度较低时（500 ℃），NiFe 合金颗粒不能够有效地俘获碳氮气体，仅能在合金颗粒表面形成较薄的氮掺杂碳壳[356-357]。图 7.10（a）～（c）给出了 500 ℃下煅烧样品的 XRD、TEM 和 HRTEM 图片，从 HRTEM 图中可以看出，样品中 NiFe 合金的表面覆盖有一层 C。在 600 ℃下，合金的溶碳能力提高，可以俘获更多的 C 并将其转化为无定形的碳纳米管[358-359]。温度升高到 700 ℃后，溶解的碳可以在合金的作用下定向生长，形成竹节状纳米管。有趣的是，样品中还可发现一些较粗的石墨烯管，这说明一些前驱体颗粒在加热过程中发生了熔合[360]。在 800 ℃下合金颗粒更容易熔合形成微米尺度的颗粒，并作为催化剂生长出石墨烯微米管。温度升高到 900 ℃后（见图 7.10（d）），合金颗粒完全熔合形成大块金属合金，不能催化形成石墨烯管。

7.3.3 NiFe-NGT 复合物的 HER 性能

由于 NiFe-NGT 复合物具有独特的结构和组分特征，该材料可用作 HER 的催化剂材料。将 NiFe-NGT-800、NiFe-NGT-700 和 NiFe-NGT-600 分别覆盖在玻碳电极上。催化剂的 HER 活性用典型的三电极系统在 0.5 M 的 H_2SO_4 溶液中测试。图 7.11（a）给出了 NiFe-NGT-800、NiFe-NGT-700 和 NiFe-NGT-600 电极的线性扫描伏安（LSV）极化曲线，商业 Pt/C 电极（20 wt%）用作对照实验。由于 Pt/C 电极具有较高的 HER 电催化活性，其过电势最小[361]。NiFe-NGT-800 电极具有更低的过电势（70.5 mV），当过电势达到 150 mV，其电流密度为 18 mA/cm^2。这一现象可能是跟氮掺杂石墨烯管在高温下产生了更多的氮掺杂结构缺陷有关。另外石墨烯 sp^2 区域增大引起的电导性增加，也是 NiFe-NGT-800 电极的 HER 电催化性能提高的重要原因。

根据 NiFe-NGT-800、NiFe-NGT-700 和 NiFe-NGT-600 电极的 LSV 极化曲线，可计算并绘制样品的 Tafel 曲线。如图 7.11（b）所示，曲线的线性区域与 Tafel 公式拟合得很好（$\eta = b \log j + a$，b 是 Tafel 斜率，j 是电流密度）。HER 电催化参数如表 7.1 所示。NiFe-NGT-800 电极的 Tafel 斜率为 63.4 mV per decade，而 NiFe-NGT-600 电极的 Tafel 斜率为 129.3 mV per decade。电极的交换电流密度也用来评价电极上电催化剂的活性，较大的交换电流密度表明催化剂具有更好的电催化活性[361-362]。NiFe-NGT-800 电极的 j_0 值为 4.05×10^{-2} mA/cm^2，是 NiFe-NGT-600 电极 j_0 值（6×10^3 mA/cm^2）的 7 倍。电极的电化学阻抗如图 7.12（a）所示，样品 NiFe-NGT-800 的半径最小，说明其阻抗最小。并且样品 NiFe-NGT-800 的电子转移阻抗 R_{ct} 为 26.8 Ω，比 NiFe-NGT-700（28.1 Ω）和 NiFe-NGT-600（32.3 Ω）小，说明该样品的电子转移速率更快。上述结果说明，NiFe-NGT-800 的 HER 催化性能最好。

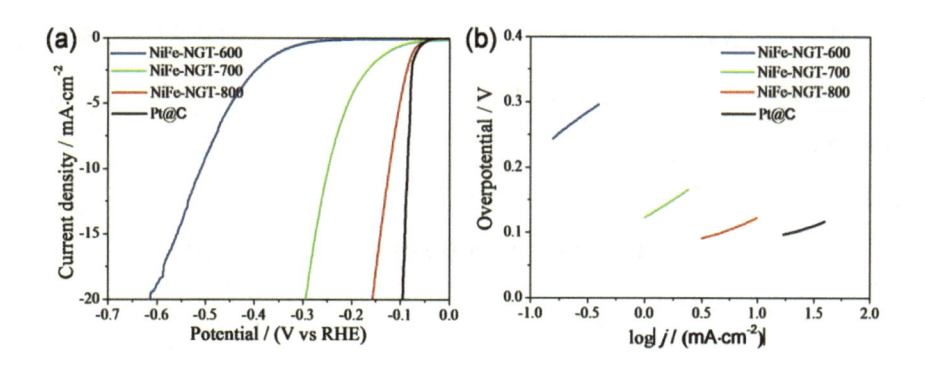

第 7 章 NiFe@N 掺杂石墨烯微管的一步法制备及其 HER 性能研究

图 7.11 （a）NiFe-NGT-800、NiFe-NGT-700、NiFe-NGT-600、Pt/C 的 LSV 极化曲线和（b）Tafel 斜率

根据经典的双电子反应模型，在酸性水介质中的阴极析氢被认为是分两个步骤发生的[349]：首先是放电步骤，在阴极的催化剂的表面一个电子转移到质子上，提供一个中间状态的氢原子结合的吸附位点（Volmer 反应：H^+（aq）$+ e^- \rightarrow H_{ads}$）；第二步是电化学沉积步骤（Heyrovsky 反应：$H_{ads} + H^+$（aq）$+ e^- \rightarrow H_2$（g））或 Tafel 重构反应（Tafel 反应：$H_{ads} + H_{ads} \rightarrow H_2$（g）），这里 H_{ads} 表示在催化剂表面吸附的氢原子。Tafel 斜率通常被用来确定一个占主导地位的反应机制[363]。根据控制步骤的不同理论算出的 Tafel 斜率有三种：116 mV（Volmer）、38 mV（Heyrovsky）和 29 mV（Tafel）[364-305]。上文提到的 NiFe-NGT-800 催化剂的 Tafel 斜率为 63.4 mV per decade，据此推测其 HER 机理为 Volmer-Heyrovsky 机制。

表 7.1 NiFe-NGT 催化剂的 HER 电催化活性

样品	过电位/mV	塔菲尔斜率/(mV dec^{-1})
NiFe-NGT-800	70.5	63.4
NiFe-NGT-700	123.5	117.7
NiFe-NGT-600	198.5	129.3

NiFe-NGT-800 的 HER 性能比文献报道的类似的 Ni、Fe 合金或 C 基材

料更优异（见表 7.2）[307, 365-367]。这证明石墨烯管与合金的复合对 HER 性能起了关键作用。一方面，石墨烯具有优异的电学性能，可以有效地提高 NiFe-NGT-800 的电导率，促进 HER 过程的电子传输和转移。另一方面，氮掺杂提高了石墨烯管中的活性位点[368-369]。

表 7.2 文献中报道的 HER 催化剂在酸性介质下的性能参数

材料	过电位/mV	塔菲尔斜率/（mV/dec^{-1}）	文献
NiMoNx/C	78	35	[307]
Co-NRCNTs	50	69	[310]
Co@BCN	96	63.7	[365]
Ni powders	300	138	[366]
Au@Zn–Fe–C	80	160	[367]

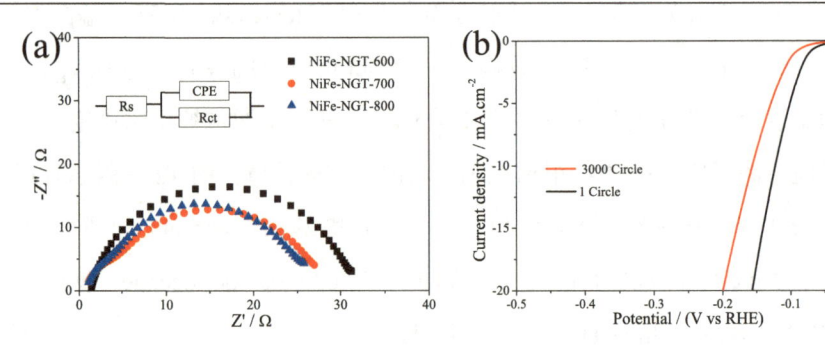

图 7.12 （a）NiFe-NGT 样品的电化学阻抗，（b）NiFe-NGT-800 在循环第一圈和第 3000 圈后的 LSV 极化曲线

氢的吸附自由能（ΔG（H_{ads}））被广泛用于评估催化材料的 HER 性能[370]。根据 ΔG（H_{ads}）绘制的火山曲线表明 ΔG（H_{ads}）趋近于零的时候可以获得最佳的 HER 性能[371]。由于 H_{ads} 产生的热力学能较高，所以纯的 CNT 或石墨烯过电势较高[108]。Bao 的课题组发现铁纳米团簇被 CNT 包覆时，由于电子从铁向邻近的碳转移，导致 CNT 费米能级附近

的电子结构可以被铁团簇改变[372]。这使得与金属相连或氮掺杂的碳原子的电子密度变大，促使生成稳定的 H_{ads}[373]。进一步证明 NiFe-NGT-800 的 HER 性能较高是由于铁镍合金与大面积的氮掺杂石墨烯管之间的协同作用引起的[361, 374-375]。

在 0.5 M H_2SO_4 溶液中对 NiFe-NGT-800 进行了 3000 次 LSV 极化测试，用于评价催化剂的稳定性，其结果如图 7.12（b）所示。结果显示 3000 个循环后，NiFe-NGT-800 电极在 η = 150 mV 电流密度由 18 mA/cm^2 降到 10 mA/cm^2。这说明 NiFe-NGT-800 电极在酸性溶液中稳定。电化学稳定性可以归因于合金颗粒和石墨烯管的化学和电子之间的耦合作用[376-378]。

7.4 本章小结

本章介绍了一种利用 NiFe 硝普盐纳米颗粒作为前驱体制备大直径氮掺杂石墨烯复合材料的方法。石墨烯管的直径和长度可以通过煅烧温度进行控制。前驱体颗粒在煅烧过程中发生了熔合，形成较大的合金颗粒，并采用一种类似于 CVD 法制备碳纳米管的过程催化碳、氮原子形成氮掺杂石墨烯管。NiFe-NGT-800 复合材料的 HER 催化性能较高，过电势和 Tafel 斜率分别为 70.5 mV 和 63.4 mV per decade，这可能是由于石墨烯管中的氮原子掺杂、石墨烯的电导以及石墨烯与 NiFe 合金颗粒之间的相互作用。该工作表明，MOF 纳米颗粒煅烧法制备的纳米结构不仅可以继承前驱体的形貌，还可以获得更加丰富的拓扑结构。

参考文献

[1] J Yuan, K Laubernds, Q Zhang, et al. Self-assembly of Microporous Manganese Oxide Octahedral Molecular Sieve Hexagonal Flakes into Mesoporous Hollow Nanospheres[J]. J. Am. Chem. Soc., 2003, 125 (17): 4966-4967.

[2] Y Zhu, Y Fang, S Kaskel. Folate-Conjugated $Fe_3O_4@SiO_2$ Hollow Mesoporous Spheres for Targeted Anticancer Drug Delivery[J]. J. Phys. Chem. C, 2010, 114 (39): 16382-16388.

[3] Y Zhu, J Shi, W Shen, et al. Stimuli-Responsive Controlled Drug Release from a Hollow Mesoporous Silica Sphere/Polyelectrolyte Multilayer Core–Shell Structure[J]. Angew. Chem., Int. Ed., 2005, 44 (32): 5083-5087.

[4] J Liu, F Liu, J Wu, et al. Recent Developments in the Chemical Synthesis of Inorganic Porous Capsules[J]. J. Mater. Chem., 2009, 19 (34): 6073-6084.

[5] J Wang, K P Loh, Y Zhong, et al. Bifunctional FePt Core−Shell and Hollow Spheres: Sonochemical Preparation and Self-assembly[J]. Chem. Mater., 2007, 19 (10): 2566-2572.

[6] R Yang, H Li, X Qiu, et al. A Spontaneous Combustion Reaction for Synthesizing Pt Hollow Capsules Using Colloidal Carbon Spheres as Templates[J]. Chem. Eur. J., 2006, 12 (15): 4083-4090.

[7] J Liu, A I Maaroof, L Wieczorek, et al. Fabrication of Hollow Metal "Nanocaps" and Their Red-Shifted Optical Absorption Spectra[J]. Adv. Mater., 2005, 17 (10): 1276-1281.

[8] R H A Ras, M Kemell, J D Wit, et al. Hollow Inorganic Nanospheres and Nanotubes with Tunable Wall Thicknesses by Atomic Layer Deposition on Self-assembled Polymeric Templates[J]. Adv. Mater., 2006, 19 (1): 1-11.

[9] S J Kim, C S Ah, D Jang. Optical Fabrication of Hollow Platinum Nanospheres by Excavating the Silver Core of Ag@Pt Nanoparticles[J]. Adv. Mater., 2007, 19 (19): 1064-1068.

[10] Z Zhong, Y Yin, B Gates, et al. Preparation of Mesoscale Hollow Spheres of TiO_2 and SnO_2 by Templating Against Crystalline Arrays of Polystyrene Beads[J]. Adv. Mater., 2000, 12 (3): 206-209.

[11] H Yoshikawa, K Hayasgida, Y Kozuka, et al. Preparation and Magnetic Properties of Hollow Nano-Spheres of Cobalt and Cobalt Oxide: Drastic Cooling-Field Effects on Remnant Magnetization of Antiferromagnet[J]. Appl. Phys. Lett., 2004, 85 (22): 5287-5289.

[12] M Agrawal, A Pich, S Gupta, et al. Synthesis of Novel Tantalum Oxide Sub-Micrometer Hollow Spheres with Tailored Shell Thickness[J]. Langmuir, 2008, 24 (3): 1013-1018.

[13] I Yamaguchi, M Watanabe, T Shinagawa, et al. Preparation of Core/Shell and Hollow Nanostructures of Cerium Oxide by Electrodeposition on a Polystyrene Sphere Template[J]. Acs Appl. Mater. Interfaces, 2009, 1 (5): 1070-1075.

[14] Z Yang, Z Niu, Y Lu, et al. Templated Synthesis of Inorganic Hollow Spheres with a Tunable Cavity Size onto Core–Shell Gel Particles[J]. Angew. Chem., Int. Ed., 2003, 42 (13): 1943-1945.

[15] M Yang, J Ma, C Zhang, et al. General Synthetic Route toward Functional Hollow Spheres with Double-Shelled Structures[J]. Angew. Chem., Int. Ed., 2005, 44 (41): 6727-6730.

[16] M Yang, J Ma, S Ding, et al. Phenolic Resin and Derived Carbon Hollow Spheres. Macromol[J]. Chem. Phys., 2006, 207 (18): 1633-1639.

[17] H Xu, W Wei, C Zhang, et al. Low-Temperature Facile Template Synthesis

of Crystalline Inorganic Composite Hollow Spheres[J]. Chem. Asian J., 2007, 2 (7): 828-836.

[18] M Chen, L Wu, S Zhou, et al. A Method for the Fabrication of Monodisperse Hollow Silica Spheres[J]. Adv. Mater., 2006, 18 (6): 801-806.

[19] X Cheng, M Chen, L Wu, et al. Novel and Facile Method for the Preparation of Monodispersed Titania Hollow Spheres[J]. Langmuir, 2006, 22 (8): 3858-3863.

[20] Z Deng, M Chen, G Gu, et al. A Facile Method to Fabricate ZnO Hollow Spheres and Their Photocatalytic Property[J]. J. Phys. Chem. B, 2008, 112 (1): 16-22.

[21] Y Liu, P Yang, W Wang, et al. Fabrication and Photoluminescence Properties of Hollow Gd_2O_3: Ln (Ln = Eu^{3+}, Sm^{3+}) Spheresvia a Sacrificial Template Method[J]. Crystengcomm, 2010, 12 (11): 3717-3723.

[22] Y Xia, R Mokaya. Hollow Spheres of Crystalline Porous Metal Oxides: a Generalized Synthesis Route Viananocasting with Mesoporous Carbon Hollow Shells[J]. J. Mater. Chem., 2005, 15 (30): 3126-3131.

[23] X Sun, Y Li. Colloidal Carbon Spheres and Their Core/Shell Structures with Noble-Metal Nanoparticles[J]. Angew. Chem., Int. Ed., 2004, 43 (5): 597-601.

[24] Y Meng, D Cheng, X Jiao. Synthesis and Characterization of $CoFe_2O_4$ Hollow Spheres[J]. Eur. J. Inorg. Chem., 2008 (25): 4019-4023.

[25] Z Chen, Z Cui, F Niu, et al. Pd Nanoparticles in Silica Hollow Spheres with Mesoporous Walls: a Nanoreactor with Extremely High Activity[J]. Chem. Commun., 2010, 46 (35): 6524-6526.

[26] Y Piao, K An, J Kim, et al. Sea Urchin Shaped Carbon Nanostructured Materials: Carbon Nanotubes Immobilized on Hollow Carbon Spheres[J]. J.

Mater. Chem., 2006, 16 (16): 2984-2989.

[27] Y Wang, G Wang, H Wang, et al. One-Pot Synthesis of Nanotube-Based Hierarchical Copper Silicate Hollow Spheres[J]. Chem. Commun., 2008, 48 (48): 6555-6557.

[28] F Piret, C Bouvy, B L Su. Monodisperse Crystalline Semiconducting ZnS Hollow Microspheres by a New Versatile Core-Shell Strategy[J]. J. Mater. Chem., 2009, 19 (33): 5964-5969.

[29] J Gao, B Zhang, X Zhang, et al. Magnetic-Dipolar-Interaction-Induced Self-assembly Affords Wires of Hollow Nanocrystals of Cobalt Selenide[J]. Angew. Chem., 2006, 118 (8): 1242-1245.

[30] Y Yin, R M Rioux, C K Erdonmez, et al. Formation of Hollow Nanocrystals Through the Nanoscale Kirkendall Effect[J]. Science, 2004, 304 (5671): 711-714.

[31] Z Shan, G Adesso, A Cabot, et al. Ultrahigh Stress and Strain in Hierarchically Structured Hollow Nanoparticles[J]. Nat. Mater., 2008, 7 (12): 947-952.

[32] A Cabot, A P Alivisatos, V F Puntes, et al. Magnetic Domains and Surface Effects in Hollow Maghemite Nanoparticles[J]. Phys. Rev. B., 2009, 797 (9): 094419-7.

[33] A Cabot, M Ibánez, P Guardia, et al. Reaction Regimes on the Synthesis of Hollow Particles by the Kirkendall Effect[J]. J. Am. Chem. Soc., 2009, 131 (32): 11326-11328.

[34] J Park, H Zheng, Y W Jun, et al. Hetero-Epitaxial Anion Exchange Yields Single-Crystalline Hollow Nanoparticles[J]. J. Am. Chem. Soc., 2009, 131 (39): 13943-13945.

[35] Y Sun, B Mayers, Y Xia. Metal Nanostructures with Hollow Interiors[J]. Adv. Mater., 2003, 15 (7-8): 641-646.

[36] B Peng, M Chen, S Zhou, et al. Fabrication of Hollow Silica Spheres Using Droplet Templates Derived from a Miniemulsion Technique[J]. J. Colloid Interface Sci., 2008, 321(1): 67-73.

[37] B P Bastakoti, S Guragain, Y Yokoyama, et al. Synthesis of Hollow $CaCO_3$ Nanospheres Templated by Micelles of Poly(Styrene-B-Acrylic Acid-B-Ethylene Glycol) in Aqueous Solutions[J]. Langmuir, 2011, 27(1): 379-384.

[38] C I Zoldesi, A Imhof. Synthesis of Monodisperse Colloidal Spheres, Capsules, and Microballoons by Emulsion Templating[J]. Adv. Mater., 2005, 17(7): 924-928.

[39] Y Zhao, J Zhang, W Li, et al. Synthesis of Uniform Hollow Silica Spheres with Ordered Mesoporous Shells in a CO_2 Induced Nanoemulsion[J]. Chem. Commun., 2009, 17(17): 2365-2367.

[40] 陈锴, 马丁, 黄伟新, 等. 利用 ostwald 熟化作用合成中空碳纳米材料[J]. 高等学校化学学报, 2008, 29(8): 1501-1504.

[41] W Cheng, K Tang, Y Qi, et al. One-Step Synthesis of Superparamagnetic Monodisperse Porous Fe_3O_4 hollow and Core-Shell Spheres[J]. J. Mater. Chem., 2010, 20(9): 1799-1805.

[42] X Wang, F Yuan, P Hu, et al. Self-Assembled Growth of Hollow Spheres with Octahedron-Like Co Nanocrystals via One-Pot Solution Fabrication[J]. J. Phys. Chem. C, 2008, 113(113): 8773-8778.

[43] X Cao, L Gu, L Zhuge, et al. Template-Free Preparation of Hollow Sb_2S_3 Microspheres as Supports for Ag Nanoparticles and Photocatalytic Properties of the Constructed Metal–Semiconductor Nanostructures[J]. Adv. Funct. Mater., 2006, 16(16): 896-902.

[44] X Dai, Y Luo, W Zhang, et al. Facile Hydrothermal Synthesis and

Photocatalytic Activity of Bismuth Tungstate Hierarchical Hollow Spheres with an Ultrahigh Surface Area[J]. Dalton Trans., 2010, 39 (14): 3426-3432.

[45] H Li, Z F Bian, J Zhu, et al. Mesoporous Titania Spheres with Tunable Chamber Stucture and Enhanced Photocatalytic Activity[J]. J. Am. Chem. Soc., 2007, 129 (27): 8406-8407.

[46] Y Wang, Q Zhu, H Zhang. Fabrication of B-Ni(OH)$_2$ and Nio Hollow Spheres by A Facile Template-Free Process[J]. Chem. Commun., 2005, 37 (5): 5231-5233.

[47] Y Wang, Q Zhu, H Zhang. Fabrication and Magnetic Properties of Hierarchical Porous Hollow Nickel Microspheres[J]. J. Mater. Chem., 2006, 16 (13): 1212-1214.

[48] J Yu, J Zhang. A Simple Template-Free Approach to TiO$_2$ Hollow Spheres with Enhanced Photocatalytic Activity[J]. Dalton Trans., 2010, 39 (25): 5860-5867.

[49] G Tian, Y Chen, W Zhou, et al. Facile Solvothermal Synthesis of Hierarchical Flower-Like Bi$_2$MoO$_6$ Hollow Spheres as High Performance Visible-Light Driven Photocatalysts[J]. J. Mater. Chem., 2011, 21 (3): 887-892.

[50] B Liu, H Zeng. Symmetric and Asymmetric Ostwald Ripening in the Fabrication of Homogeneous Core–Shell Semiconductors[J]. Small, 2005, 1 (5): 566-571.

[51] J Li, H Zeng. Hollowing Sn-Doped TiO$_2$ Nanospheres Via Ostwald Ripening[J]. J. Am. Chem. Soc., 2007, 129 (51): 15839-15847.

[52] Y Kim, J Choi, D Lee, et al. Solid-State Conversion Chemistry of Multicomponent Nanocrystals Cast in A Hollow Silica Nanosphere: Morphology-Controlled Syntheses of Hybrid Nanocrystals[J]. Acs Nano.,

2015, 9 (11) : 10719-10728.

[53] S Vaucher, M Li, S Mann. Synthesis of Prussian Blue Nanoparticles and Nanocrystal Superlattices in Reverse Microemulsions. Angew[J]. Chem. Int. Ed., 2000, 39 (10) : 1793-1796.

[54] A M Spokoyny, D Kim, A Sumrein, et al. Infinite Coordination Polymer Nano- and Microparticle Structures[J]. Chem. Soc. Rev., 2009, 38 (5) : 1218-1227.

[55] E Flügel, A Ranft, F Haase, et al. Synthetic Routes toward MOF Nanomorphologies[J]. J. Mater. Chem., 2012, 20 (22) : 10119-10133.

[56] L Catala, D Brinzei, Y Prado, et al. Core–Multishell Magnetic Coordination Nanoparticles : Toward Multifunctionality on the Nanoscale[J]. Angew. Chem. Int. Ed., 2009, 48 (1) : 1-6.

[57] X Roy, J Hui, M Rabnawaz, et al. Prussian Blue Nanocontainers : Selectively Permeable Hollow Metal-Organic Capsules from Block Ionomer Emulsion-Induced Assembly[J]. J. Am. Chem. Soc., 2011, 133 (22) : 8420-8423.

[58] K Hirai, S Furukawa, M Kondo, et al. Sequential Functionalization of Porous Coordination Polymer Crystals[J]. Angew. Chem., Int. Ed., 2011, 50 (35) : 8057-8061.

[59] S K Nune, P K Thallapally, B P McGrail, et al. A. Dohnalkova, Adsorption Kinetics in Nanoscale Porous Coordination Polymers[J]. ACS Appl. Mater. Interfaces, 2015, 7 (39) : 21712-21716.

[60] A Azadbakht, A Abbasi, N Noori. Layer-by-Layer Synthesis of Nanostructure NiBTC Porous Coordination Polymer for Iodine Removal from Wastewater[J]. J. Inorg. Organomet. P., 2016, 26 (2) : 479-487.

[61] N Shooto, C Dikio, D Wankasi, et al. Novel PVA/MOF Nanofibres :

Fabrication, Evaluation and Adsorption of Lead Ions from Aqueous Solution[J]. Nanoscale Res Lett, 2016, 11 (1): 414.

[62] W Huang, J Jiang, D Wu, et al. A Highly Stable Nanotubular MOF Rotator for Selective Adsorption of Benzene and Separation of Xylene Isomers[J]. Inorg. Chem., 2015, 54 (22): 10524-10526.

[63] L Zhang, H B Wu, S Madhavi, et al. Formation of Fe_2O_3 Microboxes with Hierarchical Shell Structures from Metal–Organic Frameworks and Their Lithium Storage Properties[J]. J. Am. Chem. Soc., 2012, 134 (42): 17388-17391.

[64] L Zhang, H B Wu, R Xu, et al. Porous Fe_2O_3 Nanocubes Derived from Mofs for Highly Reversible Lithium Storage[J]. Crystengcomm, 2013, 15 (45): 9332-9335.

[65] M Hu, A A Belik, M Imura, et al. Synthesis of Superparamagnetic Nanoporous Iron Oxide Particles with Hollow Interiors by Using Prussian Blue Coordination Polymers[J]. Chem. Mater., 2012, 24 (4): 2698-2707.

[66] L Hu, N Yan, Q W Chen, et al. Fabrication Based on the Kirkendall Effect of Co_3O_4 Porous Nanocages with Extraordinarily High Capacity for Lithium Storage[J]. Chem. Eur. J., 2012, 18 (29): 8971-8977.

[67] L Hu, P Zhang, H Zhong, et al. Foamlike Porous Spinel $Mn_xCO_{3-x}O_4$ Material Derived from $Mn_3[Co(CN)_6]_2 \cdot nH_2O$ Nanocubes: A Highly Efficient Anode Material for Lithium Batteries[J]. Chem. Eur. J., 2012, 18 (47): 15049-15056.

[68] L Hu, Y M Huang, Q W Chen. $Fe_xCO_{3-x}O_4$ Nanoporous Particles Stemmed from Metal–Organic Frameworks $Fe_3[Co(CN)_6]_2$: A Highly Efficient Material for Removal of Organic Dyes from Water[J]. J. Alloys Compd., 2013, 559: 57-63.

[69] A Banerjee, U Singh, V Aravindan, et al. Synthesis of CuO Nanostructures from Cu-Based Metal Organic Framework （MOF-199） for Application as Anode for Li-Ion Batteries[J]. Nano Energy, 2013, 2（6）: 1158-1163.

[70] L Hu, Y M Huang, F P Zhang, et al. CuO/Cu_2O Composite Hollow Polyhedrons Fabricated from Metal–Organic Framework Templates for Lithium-Ion Battery Anodes with a Long Cycling Life[J]. Nanoscale, 2013, 5（10）: 4186-4190.

[71] Y Yan, F Du, X Shen, et al. Large-Scale Facile Synthesis of Fe-Doped SnO_2 Porous Hierarchical Nanostructures and Their Enhanced Lithium Storage Properties[J]. J. Mater. Chem. A, 2014, 2（38）: 15875-15882.

[72] L Hou, L Lian, L Zhang, et al. Self-Sacrifice Template Fabrication of Hierarchical Mesoporous Bi-Component-Active ZnO/$ZnFe_2O_4$ Sub-Microcubes as Superior Anode towards High-Performance Lithium-Ion Battery[J]. Adv. Funct. Mater., 2015, 25（2）: 238-246.

[73] R Salunkhe, Y Kamachi, N L Torad, et al. Fabrication of Symmetric Supercapacitors Based on MOF-Derived Nanoporous Carbons[J]. J. Mater. Chem. A, 2014, 2（46）: 19848-19854.

[74] H Wu, B Xia, L Yu, et al. Porous Molybdenum Carbide Nano-Octahedrons Synthesized Via Confined Carburization in Metal-Organic Frameworks for Efficient Hydrogen Production[J]. Nat Commun., 2015, 6: 6512.

[75] H Lee, S Choi, M Oh. Well-Dispersed Hollow Porous Carbon Spheres Synthesized by Direct Pyrolysis of Core–Shell Type Metal–Organic Frameworks and Their Sorption Properties[J]. Chem. Commun., 2014, 50（34）: 4492-4495.

[76] L Ma, X Shen, J Zhu, et al. Co_3ZnC Core–Shell Nanoparticle Assembled Microspheres/Reduced Graphene Oxide as an Advanced Electrocatalyst for

Hydrogen Evolution Reaction in an Acidic Solution[J]. J. Mater. Chem. A,2015,3(20): 11066-11073.

[77] P Su,H Xiao,J Zhao,et al. Nitrogen-Doped Carbon Nanotubes Derived from Zn–Fe-ZIF Nanospheres and Their Application as Efficient Oxygen Reduction Electrocatalysts with in Situ Generated Iron Species[J]. Chem. Sci., 2013,4(7): 2941-2946.

[78] G Wu,P Zelenay. Nanostructured Nonprecious Metal Catalysts for Oxygen Reduction Reaction[J]. Acc. Chem. Res.,2013,46(8): 1878-1889.

[79] Q Li,P Xu,W Gao,et al. Graphene/Graphene-Tube Nanocomposites Templated from Cage-Containing Metal-Organic Frameworks for Oxygen Reduction in Li–O$_2$ Batteries[J]. Adv. Mater.,2014,26(9): 1378-1386.

[80] Q Li, H Pan, D Higgins, et al. Metal-Organic Framework-Derived Bamboo-Like Nitrogen-Doped Graphene Tubes as an Active Matrix for Hybrid Oxygen-Reduction Electrocatalysts[J]. Small,2015,11(12): 1443-1452.

[81] F Zou, Y Chen, K Liu, et al. Metal Organic Frameworks Derived Hierarchical Hollow Nio/Ni/Graphene Composites for Lithium and Sodium Storage[J]. Acs Nano.,2016,10(1): 377-386.

[82] 张金龙. 光催化[M]. 上海：华东理工大学出版社，2004.

[83] A Fujishima, K Honda. Electrochemical Photolysis of Water at a Semiconductor Electrode[J]. Nature,1972,238(238): 37-38.

[84] 李敏，崔山山. 纳米催化剂研究进展[J]. 材料导报，2006，20（S1）：8-12.

[85] 王艳芹，张莉. 掺杂过渡金属离子的 TiO$_2$ 复合纳米粒子光催化剂[J]. 高等学校化学学报，2000，21（6）：958-960.

[86] K Ikeda, J Sato, N Fujimoto, et al. Plasmonic Enhancement of Raman

Scattering on Non-Sers-Active Platinum Substrates[J]. J. Phys. Chem. C, 2009, 113 (27): 11816-11821.

[87] R Li, W Chen, H Kobayashi, et al. Platinum-Nanoparticle-Loaded Bismuth Oxide: an Efficient Plasmonic Photocatalyst Active Under Visible Light[J]. Green Chem., 2010, 12 (2): 212-215.

[88] 张莹, 龚昌杰, 燕宁宁, 等. 贵金属改性 TiO_2 光催化剂的机理及研究进展[J]. 材料导报, 2011, 25 (15): 46-49.

[89] 孙陆威. Zno 基 II 型纳米异质结构的合成与光学性质研究[D]. 杭州: 浙江大学, 2011.

[90] J Testa, M Grela, M Litter. Experimental Evidence in Favor of an Initial One-Electron-Transfer Process in the Heterogeneous Photocatalytic Reduction of Chromium (VI) Over TiO_2[J]. Langmuir, 2001, 17 (12): 3515-3517.

[91] M Kang, H Han, K Kim. Enhanced Photodecomposition of 4-Chlorophenol in Aqueous Solution by Deposition of CdS on TiO_2[J]. J. Photoch. Photobio. A: Chem., 1999, 125 (1-3): 119-125.

[92] S Khanchandani, S Kundu, A Patra. Shell Thickness Dependent Photocatalytic Properties of ZnO/CdS Core-Shell Nanorods[J]. J. Phys. Chem. C., 2012, 116 (44): 23653-23662.

[93] 张骞, 周莹, 张钊, 等. 表面等离子体光催化材料[J]. 化学进展, 2013, 25 (12): 2020-2027.

[94] P Wang, B Huang, X Qin, et al. Ag@AgCl: A Highly Efficient and Stable Photocatalyst Active Under Visible Light. Angew[J]. Chem. Int. Ed., 2008, 47 (41): 7931-7933.

[95] W Wu, S Zhang, J Zhou, et al. Controlled Synthesis of Monodisperse Sub-100 nm Hollow SnO_2 Nanospheres: A Template-and Surfactant-Free

Solution-Phase Route, the Growth Mechanism, Optical Properties, and Application as a Photocatalyst[J]. Chem. -Eur. J, 2011, 17(35): 9708-9719.

[96] K N Kim, H Jung, W Lee. Hollow cobalt ferrite–polyaniline nanofibers as magnetically separable visible-light photocatalyst for photodegradation of methyl orange[J]. J. Photoch. Photobio. A., 2016, 321: 257-265.

[97] Y Chen, W Li, J Wang, et al. Microwave-assisted ionic liquid synthesis of Ti^{3+} self-doped TiO_2 hollow nanocrystals with enhanced visible-light photoactivity[J]. Appl. Catal. B: Environ., 2016, 191: 94-105.

[98] Q Wang, L Yuan, M Dun, et al. Synthesis and characterization of visible light responsive Bi_3NbO_7 porous nanosheets photocatalyst[J]. Appl. Catal. B: Environ., 2016, 196: 127-134.

[99] H Chen, G Yu, G Li, et al. Unique Electronic Structure in a Porous Ga-In Bimetallic Oxide Nano Photocatalyst with Atomically Thin Pore Walls[J]. Angew. Chem. Int. Ed., 2016, 55(38): 11442-11446.

[100] J A Turner. A Realizable Renewable Energy Future[J]. Science, 1999, 285(5428): 687-689.

[101] Z Huang, C Wang, L Pan, et al. Enhanced Photoelectrochemical Hydrogen Production Using Silicon Nanowires@MoS_3[J]. Nano Energy, 2013, 2(6): 1337-1346.

[102] C Morales-Guio, L Stern, X Hu. Nanostructured Hydrotreating Catalysts for Electrochemical Hydrogen Evolution[J]. Chem. Soc. Rev., 2014, 43(18): 6555-6569.

[103] M S Faber, S Jin. Earth-Abundant Inorganic Electrocatalysts and their Nanostructures for Energy Conversion Applications[J]. Energy Environ. Sci., 2014, 7(11): 3519-3542.

[104] W F Chen, J T Muckerman, E Fujita. Recent Developments in Transition

Metal Carbides and Nitrides As Hydrogen Evolution Electrocatalysts[J]. Chem. Commun., 2013, 49 (79): 8896-8909.

[105] X Zou, Y Zhang. Noble Metal-Free Hydrogen Evolution Catalysts for Water Splitting[J]. Chem. Soc. Rev., 2015, 46 (36): 5148-5180.

[106] 张岐. 新生能源的挑战: 有关镍铁氢化酶研究的新进展[J]. 科学通报, 1998, 43 (9): 897-902.

[107] J F Callejas, J M Mcenaney, C G Read, et al. Electrocatalytic and Photocatalytic Hydrogen Production from Acidic and Neutral-pH Aqueous Solutions Using Iron Phosphide Nanoparticles[J]. Acs Nano, 2014, 8 (11): 11101-11107.

[108] J Deng, P Ren, D Deng, et al. Highly Active and Durable Non-Precious-Metal Catalysts Encapsulated in Carbon Nanotubes for Hydrogen Evolution Reaction[J]. Energy Environ. Sci., 2014, 7 (6): 1919-1923.

[109] L Fan, P Liu, X Yan, et al. Atomically Isolated Nickel Species Anchored on Graphitized Carbon for Efficient Hydrogen Evolution Electrocatalysis[J]. Nature Comm., 2016, 7: 10667.

[110] Y Feng, X Yu, U Paik. Nickel Cobalt Phosphides Quasi-Hollow Nanocubes as an Efficient Electrocatalyst for Hydrogen Evolution in Alkaline Solution[J]. Chem. Commun., 2016, 52 (8): 1633-1636.

[111] Y Tang, Z Jiang, G Xing, et al. Efficient Ag@AgCl Cubic Cage Photocatalysts Profit from Ultrafast Plasmon-Induced Electron Transfer Processes[J]. Adv. Funct. Mater., 2013, 23 (23): 2932-2940.

[112] Y Li, Y Ding. Porous AgCl/Ag Nanocomposites with Enhanced Visible Light Photocatalytic Properties[J]. J. Phys. Chem. C, 2010, 114 (7): 3175-3179.

[113] L Kuai, B Geng, X Chen, et al. Facile Subsequently Light-Induced Route to Highly Efficient and Stable Sunlight-Driven Ag-AgBr Plasmonic

Photocatalyst[J]. Langmuir, 2010, 26 (24): 18723-18727.

[114] D Chen, S Yoo, Q Huang, et al. Sonochemical Synthesis of Ag/AgCl Nanocubes and Their Efficient Visible-Light-Driven Photocatalytic Performance[J]. Chem Eur J., 2012, 18 (17): 5192-5200.

[115] Z Yan, G Compagnini, D B Chrisey. Generation of AgCl Cubes by Excimer Laser Ablation of Bulk Ag In Aqueous NaCl Solutions[J]. J. Phys. Chem. C, 2011, 115 (12): 5058-5062.

[116] H Wang, X Lang, J Gao, et al. Polyhedral AgBr Microcrystals with an Increased Percentage of Exposed {111} Facets as a Highly Efficient Visible-Light Photocatalyst[J]. Chem. Eur. J., 2012, 18 (15): 4620-4626.

[117] L Han, P Wang, C Zhu, et al. Facile Solvothermal Synthesis of Cube-Like Ag@AgCl: A Highly Efficient Visible Light Photocatalyst[J]. Nanoscale, 2011, 3 (7): 2931-2935.

[118] C An, S Peng, Y Sun. Facile Synthesis of Sunlight-Driven Agcl: Ag Plasmonic Nanophotocatalyst[J]. Adv. Mater., 2010, 22 (23): 2570-2574.

[119] C An, R Wang, S Wang, et al. Converting Agcl Nanocubes to Sunlight-Driven Plasmonic AgCl: Ag Nanophotocatalyst with High Activity and Durability[J]. J. Mater. Chem., 2011, 21 (31): 11532-11536.

[120] Y Bi, J Ye. In Situ Oxidation Synthesis of Ag/AgCl Core–Shell Nanowires and their Photocatalytic Properties[J]. Chem. Commun., 2009, 43 (43): 6551-6553.

[121] Y Bi, J Ye. Direct Conversion of Commercial Silver Foils Into High Aspect Ratio Agbr Nanowires with Enhanced Photocatalytic Properties[J]. Chem. Eur. J., 2010, 16 (34): 10327-10331.

[122] G Liu, J Yu, G Lu. Crystal Facet Engineering of Semiconductor Photocatalysts: Motivations, Advances and Unique Properties[J]. Chem

Commun., 2011, 47 (24): 6763-6783.

[123] M Zhu, P Chen, M Liu. Highly Efficient Visible-Light-Driven Plasmonic Photocatalysts Based on Graphene Oxide-Hybridized One-Dimensional Ag/AgCl Heteroarchitectures[J]. J. Mater. Chem., 2012, 22 (40): 21487-21494.

[124] R Dong, B Tian, C Zeng, et al. Ecofriendly Synthesis and Photocatalytic Activity of Uniform Cubic Ag@AgCl Plasmonic Photocatalyst[J]. J. Phys. Chem. C, 2013, 117 (1): 213-220.

[125] J Zheng, X Wang, W Li. Cubic Nickel Frames: One-Pot Synthesis, Magnetic Properties and Application in Water Treatment[J]. Crystengcomm, 2012, 14(22): 7616-7620.

[126] H Goesmann, C Feldmann. Nanoparticulate Functional Materials[J]. Angew. Chem., 2010, 49 (8): 1362-1395.

[127] G Dong, Y Zhu. Room-Temperature Solution Synthesis of Ag_2Te Hollow Microspheres and Dendritic Nanostructures, and Morphology Dependent Thermoelectric Properties[J]. Crystengcomm, 2012, 14 (5): 1805-1811.

[128] H Fan, U Gosele, M Zacharias. Formation of Nanotubes and Hollow Nanoparticles Based on Kirkendall and Diffusion Processes: A Review[J]. Small, 2007, 3 (10): 1660-1671.

[129] Y Yin, R Rioux, C Erdonmez, et al. Formation of Hollow Nanocrystals Through the Nanoscale Kirkendall Effect[J]. Science, 2004, 304 (5671): 711-714.

[130] Y Yin, C K Erdonmez, A Cabot, et al. Colloidal Synthesis of Hollow Cobalt Sulfide Nanocrystals[J]. Adv. Funct. Mater., 2006, 16 (11): 1389-1399.

[131] Y Wang, L Cai, Y Xia. Monodisperse Spherical Colloids of Pb and Their

Use as Chemical Templates to Produce Hollow Particles[J]. Adv. Mater., 2005, 17（4）: 473-477.

[132] H Dong, S Hughes, Y Yin, et al. Cation Exchange Reactions in Ionic Nanocrystals[J]. Science, 2004, 36（8）: 1009-1012.

[133] L Dloczik, R Konenkamp. Nanostructure Transfer in Semiconductors by Ion Exchange[J]. Nano Lett., 2003, 3（5）: 651-653.

[134] M Kovalenko, D Talapin, M Loi, et al. Quasi-Seeded Growth of Ligand-Tailored Pbse Nanocrystals Through Cation-Exchange- Mediated Nucleation[J]. Angew. Chem. Int. Ed., 2008, 47（16）: 3029-3033.

[135] J Park, H Zheng, A P Alivisatos. Hetero-Epitaxial Anion Exchange Yields Single-Crystalline Hollow Nanoparticles[J]. J. Am. Chem. Soc., 2009, 131（39）: 13943-13945.

[136] J Eckert Jr., C Hung-Houston, B Gersten, et al. Kinetics and Mechanisms of Hydrothermal Synthesis of Barium Titanate[J]. J. Am. Ceram. Soc., 1996, 79（11）: 2929-2939.

[137] Z Lou, B Huang, X Qin. One-Step Synthesis of Agcl Concave Cubes by Preferential Overgrowth Along 〈111〉 and 〈110〉 Directions[J]. Chem. Commun., 2012, 48（29）: 3488-3490.

[138] R Brenier. Silver Nanoparticle Oxide Coating via A Surface-Initiated Reduction Process[J]. J. Phys. Chem. C, 2009, 113（5）: 1758-1763.

[139] J Kim. Reduction of Silver Nitrate In Ethanol by Poly（N-Vinylpyrrolidone）[J]. J. Ind. Eng. Chem., 2007, 13（4）: 566-570.

[140] K Doll, N M Harrison. Theoretical Study of Chlorine Adsorption on the Ag （111） Surface[J]. Phys. Rev. B, 2001, 63（16）: 165410.

[141] M S Bootharaju, G K Deepesh, T Udayabhaskararao, et al. Atomically Precise Silver Clusters for Efficient Chlorocarbon Degradation[J]. J. Mater.

Chem. A，2012，1（3）：611-620.

[142] M Moskovits， B Vlckova. Adsorbate-Induced Silver Nanoparticle Aggregation Kinetics[J]. J. Phys. Chem. B.，2005，109（31）：14755-14758.

[143] S Wu，C Chen，X Shen，et al. One-Pot Synthesis，Formation Mechanism and Nearinfrared Fluorescent Properties of Hollow and Porous A-Mercury Sulfide[J]. Crystengcomm，2013，15（20）：4162-4166.

[144] Y Wang，D Wan，S Xie，et al. Synthesis of Silver Octahedra with Controlled Sizes and Optical Properties via Seed-Mediated Growth[J]. Acsnano.，2013，7（5）：4586-4594.

[145] J Jiang，L Zhang. Rapid Microwave-assisted Nonaqueous Synthesis and Growth Mechanism of AgCl/Ag，and Its Daylight-Driven Plasmonic Photocatalysis[J]. Chem. Eur. J，2011，17（13）：3710-3717.

[146] H Xu，H Li，J Xia，et al. One-Pot Synthesis of Visible-Light-Driven Plasmonic Photocatalyst Ag/AgCl in Ionic Liquid[J]. ACS Appl._Mater. Inter，2011，3（1）：22-29.

[147] Y Zhang，Z Tang，X Fu，et al. TiO_2−Graphene Nanocomposites for Gas-Phase Photocatalytic Degradation of Volatile Aromatic Pollutant：Is TiO_2−Graphene Truly Different from Other TiO_2−Carbon Composite Materials?[J]. ACS Nano，2010，4（12）：7303-7314.

[148] N Zhang，Y H Zhang，Y J Xu. Recent Progress on Graphene-Based Photocatalysts：Current Status and Future Perspectives[J]. Nanoscale，2012，4（19）：5792-5813.

[149] M Choi，K Shin，J Jang. Plasmonic Photocatalytic System Using Silver Chloride/Silver Nanostructures Under Visible Light[J]. J. Colloid Interface Sci.，2010，341（1）：83-87.

[150] L Ji，W Chen，L Duan. Mechanisms for Strong Adsorption of Tetracycline

to Carbon Nanotubes: A Comparative Study Using Activated Carbon and Graphite as Adsorbents[J]. Environ. Sci. Technol., 2009, 43 (7): 2322-2327.

[151] P H Chang, Z H Li, T L Yu, et al. Sorptive Removal of Tetracycline from Water by Palygorskite[J]. J. Hazard. Mater., 2009, 165 (1-3): 148-155.

[152] J Kang, H Liu, Y M Zheng, et al. Application of Nuclear Magnetic Resonance Spectroscopy, Fourier Transform Infrared Spectroscopy, Uv-Visible Spectroscopy and Kinetic Modeling for Elucidation of Adsorption Chemistry in Uptake of Tetracycline by Zeolite Beta[J]. J. Colloid Interface Sci., 2011, 354 (1): 261-267.

[153] X Zhang, Z Ai, F Jia, et al. Generalized One-Pot Synthesis, Characterization, and Photocatalytic Activity of Hierarchical BiOX (X = Cl, Br, I) Nanoplate Microspheres[J]. J. Phys. Chem. C, 2008, 112 (3): 747-753.

[154] X Xiao, W D Zhang. Facile Synthesis of Nanostructured Bioi Microspheres with High Visible Light-Induced Photocatalytic Activity[J]. J. Mater. Chem., 2010, 20 (28): 5866-5870.

[155] W Li, D Li, Y Lin, et al. Evidence for the Active Species Involved in the Photodegradation Process of Methyl Orange on TiO_2[J]. J. Phys. Chem. C, 2012, 116 (5): 3552-3560.

[156] X F Yang, H Y Cui, Y Li, et al. Fabrication of Ag_3PO_4-Graphene Composites with Highly Efficient and Stable Visible Light Photocatalytic Performance[J]. ACS Catal. 2013, 3 (3): 363-369.

[157] G Z Liao, S Chen, X Quan, et al. Graphene Oxide Modified G-C_3N_4 Hybrid with Enhanced Photocatalytic Capability Under Visible Light Irradiation[J]. J. Mater. Chem., 2012, 22 (6): 2721-2726.

[158] M R Hoffmann,S T Martin,W Choi,et al. Environmental Applications of Semiconductor Photocatalysis[J]. Chem. Rev. 1995,95(1):69-96.

[159] B Tian, J Zhang. Morphology-Controlled Synthesis and Applications of Silver Halide Photocatalytic Materials[J]. Catal Surv Asia,2012,16(4):210-230.

[160] S S Soni,M J Henderson,J F Bardeau,et al. Visible-Light Photocatalysis in Titania-Based Mesoporous Thin Films[J]. Adv. Mater.,2008,20(8):1493-1498.

[161] Y Zhong,J Wang,R Zhang,et al. Morphology-Controlled Self-assembly and Synthesis of Photocatalytic Nanocrystals[J]. Nano Lett.,2014,14(12):7175-7179.

[162] C Burd, X Chen, R Narayanan, et al. Chemistry and Properties of Nanocrystals of Different Shapes[J]. Chem. Rev.,2005,105(4):1025-1102.

[163] X Xia,J Zeng,L Oetjen,et al. Quantitative Analysis of the Role Played by Poly(Vinylpyrrolidone)in Seed-Mediated Growth of Ag Nanocrystals[J]. J. Am. Chem. Soc.,2012,134(3):1793-1801.

[164] M Pang,A J Cairns,Y Liu,et al. Highly Monodisperse M^{III}-Based Soc-MOFs (M=In and Ga)with Cubic and Truncated Cubic Morphologies[J]. J. Am. Chem. Soc.,2012,134(32):13176-13179.

[165] X Mou,Y Li,B Zhang,et al. Crystal Phase and Morphology Controlled Synthesis of Fe_2O_3 Nanomaterials[J]. Eur. J. Inorg. Chem.,2012,2012(16):2684-2690.

[166] J Kundu, D Pradhan. Controlled Synthesis and Catalytic Activity of Copper Sulfide Nanostructured Assemblies with Different Morphologies[J].

Acs Appl. Mater. Inter., 2014, 6（3）: 1823-1834.

[167] H Zhang, Y Lu, H Liu. One-Pot Synthesis of High-Index Faceted AgCl Nanocrystals with Trapezohedral, Concave Hexoctahedral Structures and Their Photocatalytic Activity[J]. Nanoscale, 2015, 7（27）: 11591-11601.

[168] T K Sau, A L Rogach. Nonspherical Noble Metal Nanoparticles: Colloid Chemical Synthesis and Morphology Control[J]. Adv. Mater., 2010, 22（16）: 1781-1804.

[169] M Pal, J Garcia Serrano, P Santiago, et al. Size-Controlled Synthesis of Spherical TiO_2 Nanoparticles: Morphology, Crystallization, and Phase Transition[J]. J. Phys. Chem. C, 2007, 111（1）: 96-102.

[170] Y Sun. Conversion of Ag Nanowires to Agcl Nanowires Decorated with Au Nanoparticles and Their Photocatalytic Activity[J]. J Phys. Chem. C, 2010, 114（5）: 2127-2133.

[171] Z Lin, J Xiao, J Yan, et al. Ag/AgCl Plasmonic Cubes with Ultrahigh Activity as Advanced Visible-Light Photocatalysts for Photodegrading Dyes[J]. J. Mater. Chem. A, 2015, 3（14）: 7649-7658.

[172] H Inoue, T Nakazawa, T Mitsuhashi, et al. Characterization of Prussian Blue and Its Thermal Decomposition Products[J]. Hyperfine Interact, 1989, 46（1）: 723-731.

[173] E V Shevchenko, D M Talapin, H Schnablegger, et al. Study of Nucleation and Growth in the Organometallic Synthesis of Magnetic Alloy Nanocrystals: the Role of Nucleation Rate in Size Control of $CoPt_3$ Nanocrystals[J]. J. Am. Chem. Soc., 2003, 125（30）: 9090-9101.

[174] S Auer, D Frenkel. Prediction of Absolute Crystal-Nucleation Rate in Hard-Sphere Colloids[J]. Nature, 2001, 409（6823）: 1020-1023.

[175] X Yan, D Xu, D Xue, SO_4^{2-} Ions Direct the One-Dimensional Growth of $5mg(OH)_2 \cdot MgSO_4 \cdot 2H_2O$[J]. Acta Mater., 2007, 55（17）: 5747-5757.

[176] D Xu, D Xue. Chemical Bond Analysis of the Crystal Growth of KDP and ADP[J]. J. Cryst. Growth, 2006, 286（1）: 108-113.

[177] Y Zheng, W Liu, T Lv, et al. Seed-Mediated Synthesis of Gold Tetrahedra in High Purity and with Tunable, Well-Controlled Sizes[J]. Chem. Asian J. 2014, 9（9）: 2635-2640.

[178] Y Wang, S Xie, J Liu, et al. Shape-Controlled Synthesis of Palladium Nanocrystals: A Mechanistic Understanding of the Evolution from Octahedrons to Tetrahedrons[J]. Nano Lett. 2013, 13（5）: 2276-2281.

[179] H Zhang, Y Lu, H Liu, et al. Controllable Synthesis of Three-Dimensional Branched Gold Nanocrystals Assisted by Cationic Surfactant Poly（Diallyldimethylammonium） Chloride in Acidic Aqueous Solution[J]. RSC Adv., 2014, 4（69）: 36757-36764.

[180] M Zayat, D Einot, R Reisfeld. In-Situ Formation of Agcl Nanocrystallites in Films Prepared by the Sol-Gel and Silver Nanoparticles in Silica Glass Films[J]. J. Sol-Gel Sci. Technol., 1997, 10（1）: 67-74.

[181] C D Wagner, W M Riggs, L E Davis, et al. Muilenberg in Handbook of X-Ray Photoelectron Spectroscopy, Physical Electronics Division[J]. Perkin-Elmer Corp., Eden Prairie, 1979.

[182] V Djoković, R Krsmanović, D K Božanić, et al. Adsorption of Sulfur onto A Surface of Silver Nanoparticles Stabilized with Sago Starch Biopolymer[J]. Colloids Surf. B, 2009, 73（1）: 30-35.

[183] P Gangopadhyay, R Kesavamoorthy, S Bera, et al. Optical Absorption and Photoluminescence Spectroscopy of the Growth of Silver Nanoparticles[J].

Phys. Rev. Lett., 2005, 94 (4): 47403.

[184] H Yu, S Ouyang, S Yan, et al. Sol-Gel Hydrothermal Synthesis of Visible-Light-Driven Cr-Doped $SrTiO_3$ for Efficient Hydrogen Production[J]. J. Mater. Chem., 2011, 21 (30): 11347-11351.

[185] X Zong, H Yan, G Wu, et al. Enhancement of Photocatalytic H_2 Evolution on Cds by Loading MoS_2 As Cocatalyst Under Visible Light Irradiation[J]. J. Am. Chem. Soc., 2008, 130 (23): 7176-7177.

[186] N Liédana, E Marín, C Téllez, et al. One-Step Encapsulation of Caffeine in SBA-15 Type and Non-Ordered Silicas[J]. Chem. Eng. J. 2013, 223(5): 714-721.

[187] J Muñoz-Pallares, A Corma, J Primo, et al. Zeolites as Pheromone Dispensers[J]. J. Agric. Food Chem. 2001, 49 (10): 4801-4807.

[188] J Li, Z Shao, C Chen, et al. Hierarchical GOs/Fe_3O_4/PANI Magnetic Composites as Adsorbent for Ionic Dye Pollution Treatment[J]. RSC Adv. 2014, 4 (72): 38192-38198.

[189] J Hu, D Shao, C Chen, et al. Removal of 1-Naphthylamine from Aqueous Solution by Multiwall Carbon Nanotubes/Iron Oxides/Cyclodextrin Composite[J]. J. Hazard. Mater. 2011, 185 (1): 463-471.

[190] D Zhao, X Yang, C Chen, et al. Enhanced Photocatalytic Degradation of Methylene Blue on Multiwalled Carbon Nanotubes–TiO_2[J]. J. Colloid Interface Sci, 2013, 398 (19): 234-239.

[191] Y Zhao, H Chen, J Li, et al. Hierarchical MWCNTs/Fe_3O_4/PANI Magnetic Composite as Adsorbent for Methyl Orange Removal[J]. J. Colloid Interface Sci., 2015, 450: 189-195.

[192] R J Kuppler, D J Timmons, Q Fang, et al. Potential Applications of Metal-Organic Frameworks[J]. Coord. Chem. Rev., 2009, 253 (23-24):

3042-3066.

[193] J Li，J Sculley，H Zhou. Metal-Organic Frameworks for Separations[J].Chem. Rev.，2012，112（2）：869-932.

[194] W Lu，Z Wei，Z Y Gu，et al. Tuning the Structure and Function of Metal-Organic Frameworks via Linker Design[J].Chem. Soc. Rev.，2014，43（16）：5561-5593.

[195] J Juan-Alcaniz，R Gielisse，A B Lago，et al. Towards Acid MOFs-Catalytic Performance of Sulfonic Acid Functionalized Architectures[J]. Catal. Sci. Technol.，2013，3（9）：2311-2318.

[196] Y Hwang，D Hong，J Chang，et al. Amine Grafting on Coordinatively Unsaturated Metal Centers of Mofs：Consequences for Catalysis and Metal Encapsulation[J]. Angew. Chem. Int. Ed.，2008，47（22）：4144-4148.

[197] J Evans，C Sumby，C Doonan. Post-Synthetic Metalation of Metal-Organic Frameworks[J]. Chem. Soc. Rev.，2014，43（16）：5933-5951.

[198] S Jhung，N Khan，Z Hasan. Analogous Porous Metal-Organic Frameworks：Synthesis，Stability and Application in Adsorption[J]. Crysengcomm，2012，43（14）：7099-7109.

[199] W Zhang，R Xiong. Ferroelectric Metal-Organic Frameworks[J]. Chem. Rev.，2012，112（2）：1163-11195.

[200] H Furukawa，N Ko，Y Go，et al. Ultrahigh Porosity in Metal-Organic Frameworks[J]. Science，2010，329（5990）：424-428.

[201] K Konstas，T Osl，X Yang，et al. Methane Storage in Metal Organic Frameworks[J]. J. Mater. Chem. 2012，22：16698-16708.

[202] H Wu，H Gong，D H Olson，et al. Commensurate Adsorption of Hydrocarbons and Alcohols in Microporous Metal Organic Frameworks[J]. Chem. Rev.，2012，112（2）：836-868.

[203] W Bloch, R Babarao, M Hill, et al. Post-Synthetic Structural Processing in a Metal-Organic Framework Material as a Mechanism for Exceptional CO_2/N_2 Selectivity[J]. J. Am. Chem. Soc., 2013, 135 (28): 10441-10448.

[204] A Carné-Sánchez, I Imaz, M Cano-Sarabia, et al. A Spray-Drying Strategy for Synthesis of Nanoscale Metal–Organic Frameworks and Their Assembly into Hollow Superstructures[J]. Nat. Chem., 2013, 5 (3): 203-211.

[205] M K Suh, H J Park, T K Prasad, et al. Hydrogen Storage in Metal-Organic Frameworks[J]. Chem. Rev., 2012, 112 (2): 782-835.

[206] M Yoon, R Srirambalaji, K Kim. Homochiral Metal-Organic Frameworks for Asymmetric Heterogeneous Catalysis[J]. Chem. Rev., 2012, 112(112): 1196-1231.

[207] T Lee, H L Lee, M H Tsai, et al. A Biomimetic tongue by Photoluminescent Metal–Organic Frameworks[J]. Biosens Bioelectron, 2013, 43 (19): 56-62.

[208] L Sun, H Xing, Z Liang, et al. 4 + 4 Strategy for Synthesis of Zeolitic Metal-Organic Frameworks: An Indium-Mof with SOD topology as a Light-Harvesting Antenna[J]. Chem. Commun., 2013, 49 (95): 11155-11157.

[209] N Khan, Z Hasan, S Jhung. Adsorptive Removal of Hazardous Materials Using Metal-Organic Frameworks (MOFs): A Review[J]. J. Hazard. Mater., 2013, 244-245 (2): 444-456.

[210] K Cyshosz, A Wong-Foy, A Matzger. Liquid Phase Adsorption by Micro-Porous Coordination Polymers: Removal of Organosulfur Compounds[J]. J. Am. Chem. Soc., 2008, 130 (22): 6938-6939.

[211] N A Khan, S H Jhung. Effect of Central Metal Ions of Analogous

Metal-Organic Frameworks on the Adsorptive Removal of Benzothiophene from Amodel Fuel[J]. J. Hazard. Mater., 2013, 260 (18): 1050-1056.

[212] I Ahmed, N Khan, S Jhung. Graphite Oxide/Metal-Organic Framework (MIL-101): Remarkable Performance in the Adsorptive Denitrogenation of Model Fuels[J]. Inorg. Chem., 2013, 52 (24): 14155-14161.

[213] F Pu, X Liu, B L Xu, et al. Miniaturization of Metal-Biomolecule Frameworks Based on Stereoselective Self-assembly and Potential Application in Water Treatment and as Antibacterial Agents[J]. Chem. Eur. J., 2012, 18 (14): 4322-4328.

[214] H Deng, S Grunder, K Cordova, et al. Large-Pore Apertures in a Series of Metal-Organic Frameworks[J]. Science, 2012, 336 (6084): 1018-1023.

[215] X Zhao, X Bu, T Wu, et al. Selective Anion Exchange with Nanogated Isoreticular Positive Metal-Organic Frameworks[J]. Nat Commun., 2013, 4: 2344.

[216] J Qin, S Zhang, D Du, et al. A Microporous Anionic Metal-Organic Framework for Sensing Luminescence of Lanthanide (Iii) Ions and Selective Absorption of Dyes by Ionic Exchange[J]. Chem. Eur. J., 2014, 20 (19): 5625-5630.

[217] Y He, J Yang, W Kan, et al. A New Microporous Anionic Metal–Organic Framework As A Platform for Highly Selective Adsorption and Separation of Organic Dyes[J]. J. Mater. Chem. A, 2015, 3 (4): 1675-1681.

[218] Z Hasan, S H Jhung. Removal of Hazardous Organics from Water Using Metal-Organic Frameworks (MOFs): Plausible Mechanisms for Selective Adsorptions[J]. J. Hazard. Mater., 2015, 283: 329-339.

[219] D Maspoch, D Ruiz-Molina, J Veciana. Old Materials with New Tricks: Multifunctional Open-Framework Materials[J]. Chem. Soc. Rev., 2007,

36（5）：770-818.

[220] M Shatruk, C Avendano, K R Dunbar. Cyanide-Bridged Complexes of Transition Metals: A Molecular Magnetism Perspective[J]. Prog. Inorg. Chem., 2009, 56: 155-334.

[221] C Yuan, A Yuan, W Liu, et al. Synthesis, Structure and Porous Properties of Prussian Blue Analogue Coordination Polymer KCd[Cr(CN)$_6$]·H$_2$O[J]. Acta. Chimica. Sinica., 2008, 66（24）: 2700-2704.

[222] X Shen, S Wu, Y Liu, et al. Morphology Syntheses and Properties of Well-Defined Prussian Blue Nanocrystals by A Facile Solution Approach[J]. J. Colloid Interface Sci., 2009, 329（1）: 188-195.

[223] G Zhu, Y Xiao, X Shen, et al. Shape and Size Tunable Synthesis of Coordination Polymer Mn$_2$W(CN)$_8$·xH$_2$O Microcrystals Through A Simple Solution Chemical Route[J]. Eur. J. Inorg. Chem., 2013, 2013（30）: 5297-5302.

[224] O N Risset, D R Talham. Effects of Lattice Misfit on the Growth of Coordination Polymer Heterostructures[J]. Chem. Mater., 2015, 27（11）: 3838-3843.

[225] H Lee, Y I Kim, J K Park, et al. Sodium Zinc Hexacyanoferrate with a Well-Defined Open Framework as a Positive Electrode for Sodium Ion Batteries[J]. Chem. Commun., 2012, 48（67）: 8416-8418.

[226] R Srivastava, D Srinivas, P Ratnasamy. Fe–Zn Double-Metal Cyanide Complexes as Novel, Solid Transesterification Catalysts[J]. J. Catal., 2006, 241（1）: 34-44.

[227] Y Yang, C Brownell, N Sadrieh, et al. Quantitative Measurement of Cyanide Released from Prussian Blue[J]. Clin. Toxicol., 2007, 45（7）: 776-781.

[228] M S Kandanapitiye, F J Wang, B Valley, et al. Selective Ion Exchange Governed by the Irving-Williams Series In $K_2Zn_3[Fe(CN)_6]_2$ Nanoparticles: toward A Designer Prodrug for Wilson'S Disease[J]. Inorg. Chem., 2015, 54(4): 1212-1214.

[229] P Gravereau, E Garnier. Structure De La Phase Cubique De I'Hexacyanoferrate(Lll) De Zinc: $Zn_3[Fe(CN)_6]_2 \cdot n H_2O$[J]. Acta Cryst. C, 1984, 40(8): 1306-1309.

[230] H Ming, N L K Torad, Y Chiang, et al. Size- and Shape-Controlled Synthesis of Prussian Blue Nanoparticles by A Polyvinylpyrrolidone-assisted Crystallization Process[J]. Crystengcomm, 2012, 14(10): 3387-3396.

[231] T Bolam, W Taylor. Colloidal Prussian Blue[J]. Trans. Faraday Soc., 1939, 35: 268-276.

[232] C Du, F Bu, D Jiang, et al. Prussian Blue Analogue $K_2Zn_3[Fe(CN)_6]_2$ Quasi Square Microplates: Large-Scale Synthesis and Their Thermal Conversion into A Magnetic Nanoporous $ZnFe_{2-x}O_4$-ZnO Composite[J]. Crystengcomm, 2013, 15(48): 10597-10603.

[233] Y Chiang, M Hu, Y Kamachi, et al. Rational Design and Synthesis of Cyano-Bridged Coordination Polymers with Precise Control of Particle Size from 20 to 500 nm[J]. Eur. J. Inorg. Chem., 2013, 2013(18): 3141-3145.

[234] S M Lee, S N Cho, J Cheon. Anisotropic Shape Control of Colloidal Inorganic Nanocrystals[J]. Adv. Mater., 2003, 15(15): 441-444.

[235] V Lamer, R Dinegar. Theory, Production and Mechanism of Formation of Monodispersed Hydrosols[J]. J. Am. Chem. Soc., 1950, 72(11): 4847-4854.

[236] X Zheng, Q Kuang, T Xu, et al. Growth of Prussian Blue Microcubes

Under A Hydrothermal Condition: Possible Nonclassical Crystallization by A Mesoscale Self-assembly[J]. J. Phys. Chem. C, 2007, 111 (12): 4499-4502.

[237] H Cölfen, M Antonietti. Mesocrystals: Inorganic Superstructures Made by Highly Parallel Crystallization and Controlled Alignment[J]. Angew. Chem. Int. Ed., 2005, 44 (35): 5576-5591.

[238] S Wu, X Shen, B Cao, et al. Shape- and Size-Controlled Synthesis of Coordination Polymer $\{[Cu(en)_2][KFe(CN)_6]\}_n$ Nano/Micro-Crystals[J]. J. Mater. Sci., 2009, 44 (23): 6447-6450.

[239] M Hu, J Jiang, Y Zeng. Prussian Blue Microcrystals Prepared by Selective Etching and Their Conversion to Mesoporous Magnetic Iron(III)Oxides[J]. Chem. Commun., 2010, 46 (7): 1133-1135.

[240] Y Wang, Y Yang, X Hao, et al. pH-Controlled Morphological Structure and Electrochemical Performances of Polyaniline/Nickel Hexacyanoferrate Nanogranules During Electrochemical Deposition[J]. J. Solid. State. Electr., 2014, 18 (10): 2885-2892.

[241] Z L Wang. Transmission Electron Microscopy of Shape-Controlled Nanocrystals and Their Assemblies[J]. J. Phys. Chem. B, 2000, 104 (6): 1153-1175.

[242] D Du, M Cao, X He, et al. Morphology-Controllable Synthesis of Microporous Prussian Blue Analogue $Zn_3[Co(CN)_6]_2 \cdot xH_2O$ Microstructures[J]. Langmuir, 2009, 25 (12): 7057-7062.

[243] J Feng, Y Wang, L Zou, et al. Synthesis of Magnetic $ZnO/ZnFe_2O_4$ by A Microwave Combustion Method, and Its High Rate of Adsorption of Methylene Blue[J]. J. Colloid Inter. Sci. 2015, 438: 318-322.

[244] F Tan, M Liu, K Li, et al. Facile Synthesis of Size-Controlled MIL-100

（Fe）with Excellent Adsorption Capacity for Methylene Blue[J]. Chem. Eng. J., 2015, 281: 360-367.

[245] E Haque, J Jun, S Jhung. Adsorptive Removal of Methyl Orange and Methylene Blue from Aqueous Solution with a Metal-Organic Framework Material, Iron Terephthalate（MOF-235）[J]. J. Hazard. Mater., 2011, 185（1）: 507-511.

[246] L Li, X Liu, H Geng, et al. A MOF/Graphite Oxide Hybrid（MOF: HKUST-1）Material for the Adsorption ofmethylene Blue from Aqueous Solution[J]. J. Mater. Chem. A, 2013, 1（35）: 10292-10299.

[247] F Luo, X Li, G He, et al. Preparation of Double-Shelled C/SiO_2 Hollow Spheres with Enhanced Adsorption Capacity[J]. Ind. Eng. Chem. Res., 2015, 54（2）: 641-648.

[248] A Yan, S Yao, Y Li, et al. Incorporating Polyoxometalates into A Porous MOF Greatly Improves Its Selective Adsorption of Cationic Dyes[J]. Chem.-Eur. J., 2014, 20（23）: 6927-6933.

[249] X Yuan, X Shi, S Zeng, et al. Activated Carbons Prepared from Biogas Residue: Characterization and Methylene Blue Adsorption Capacity[J]. J. Chem. Technol. Biot., 2011, 86（3）: 361-366.

[250] X Zhao, S Liu, Z Tang, et al. Synthesis of Magnetic Metal-Organic Framework（MOF）for Efficient Removal of Organic Dyes from Water[J]. Scientific Reports, 2015, 5: 11849.

[251] S Huo, X Yan. Metal-Organic Framework MIL-100（Fe）for the Adsorption of Malachite Green from Aqueous Solution[J]. J. Mater. Chem., 2012, 22（15）: 7449-7455.

[252] X Cao, L Gu, X Lan, et al. Spinel $ZnFe_2O_4$ Nanoplates Embedded with Ag Clusters: Preparation, Characterization, and Photocatalytic Application[J].

Mater. Chem. Phys., 2007, 106（2-3）：175-180.

[253] R Shao, L Sun, L Tang, et al. Preparation and Characterization of Magnetic Core-Shell ZnFe$_2$O$_4$@ZnO Nanoparticles and Their Application for the Photodegradation of Methylene Blue[J]. Chem. Eng. J., 2013, 217（2）：185-191.

[254] Y Fu, X Wang. Magnetically Separable ZnFe$_2$O$_4$–Graphene Catalyst and Its High Photocatalytic Performance Under Visible Light Irradiation[J]. Ind. Eng. Chem. Res., 2011, 50（12）：7210-7218.

[255] Y Sun, W Wang, L Zhang, et al. Magnetic ZnFe$_2$O$_4$ Octahedra: Synthesis and Visible Light Induced Photocatalytic Activities[J]. Mater. Lett., 2013, 98（5）：124-127.

[256] X Li, Y Hou, Q Zhao, et al. Capability of Novel ZnFe$_2$O$_4$ Nanotube Arrays for Visible-Light Induced Degradation of 4-Chlorophenol[J]. Chemosphere, 2011, 82（4）：581-586.

[257] H Lv, L Ma, P Zeng, et al. Synthesis of Floriated ZnFe$_2$O$_4$ with Porous Nanorod Structures and Its Photocatalytic Hydrogen Production Under Visible Light[J]. J. Mater. Chem., 2010, 20（20）：3665-3672.

[258] X Chen, S Shen, L Guo, et al. Semiconductor-Based Photocatalytic Hydrogen Generation. Chemical Reviews[J]. Chem. Rev., 2010, 110（11）：6503-6570.

[259] J Di, J Xia, Y Ge, et al. Novel Visible-Light-Driven CQDs/Bi$_2$WO$_6$ Hybrid Materials with Enhanced Photocatalytic Activity toward Organic Pollutants Degradation and Mechanism Insight[J]. Appl. Catal. B: Environ., 2015, 168：51-61.

[260] C Chen, Y Liang, W Zhang. ZnFe$_2$O$_4$/MWCNTs Composite with Enhanced Photocatalytic Activity Under Visible-Light Irradiation[J]. J. Alloys

Compd., 2010, 501 (1): 168-172.

[261] Y Yao, J Qin, Y Cai, et al. Facile Synthesis of Magnetic $ZnFe_2O_4$–Reduced Graphene Oxide Hybrid and Its Photo-Fenton-Like Behavior Under Visible Irradiation[J]. Environ. Sci. Pollut. R., 2014, 21 (12): 7296-7306.

[262] Y Hou, X Li, Q Zhao, et al. Electrochemical Method for Synthesis of A $ZnFe_2O_4/TiO_2$ Composite Nanotube Array Modified Electrode with Enhanced Photoelectrochemical Activity[J]. Adv. Funct. Mater., 2010, 20 (13): 2165-2174.

[263] L Sun, R Shao, L Tang, et al. Synthesis of $ZnFe_2O_4$/ZnO Nanocomposites Immobilized on Graphene with Enhanced Photocatalytic Activity Under Solar Light Irradiation[J]. J. Alloys Compd., 2013, 564 (7): 55-62.

[264] X Guo, H Zhu, Q Li. Visible-Light-Driven Photocatalytic Properties of $ZnO/ZnFe_2O_4$ Core/Shell Nanocablearrays[J]. Appl. Catal. B: Environ., 2014, 160-161 (1): 408-414.

[265] I Arabatzis, T Stergiopoulos, M Bernard, et al. Silver-Modified Titanium Dioxide Thin Films for Efficient Photodegradation of Methyl Orange[J]. Appl. Catal. B: Environ., 2003, 42 (2): 187-201.

[266] R Georgekutty, M K Seery, S C Pillai. A Highly Efficient Ag-ZnO Photocatalyst: Synthesis, Properties, and Mechanism[J]. J. Phys. Chem. C, 2008, 112 (35): 13563-13570.

[267] L Zhang, H Wu, X Lou. Metal–Organic-Frameworks-Derived General Formation of Hollow Structures with High Complexity[J]. J. Am. Chem. Soc., 2013, 135 (29): 10664-10672.

[268] L Hu, Q Chen. Hollow/Porous Nanostructures Derived from Nanoscale Metal–Organic Frameworks towards High Performance Anodes for Lithium-Ion Batteries[J]. Nanoscale, 2014, 6 (3): 1236-1257.

[269] Z Wang, X Li, H Xu, et al. Porous Anatase TiO_2 Constructed from a Metal–Organic Framework for Advanced Lithium-Ion Battery Anodes[J]. J. Mater. Chem. A, 2014, 2 (31): 12571-12575.

[270] C Du, F Bu, D Jiang, et al. Prussian Blue Analogue $K_2Zn_3[Fe(CN)_6]_2$ Quasi Square Microplates: Large-Scale Synthesis and Their Thermal Conversion into A Magnetic Nanoporous $ZnFe_{2-x}O_4$–ZnO Composite[J]. Crystengcomm, 2013, 15 (48): 10597-10603.

[271] T Kim, K Lee, J Cheon, et al. Nanoporous Metal Oxides with Tunable and Nanocrystalline Frameworks Via Conversion of Metal–Organic Frameworks[J]. J. Am. Chem. Soc., 2013, 135 (24): 8940-8946.

[272] Y Yan, F Du, X Shen, et al. Large-Scale Facile Synthesis of Fe-Doped SnO_2 Porous Hierarchical Nanostructures and Their Enhanced Lithium Storage Properties[J]. J. Mater. Chem. A, 2014, 2 (38): 15875-15882.

[273] B Kong, J Tang, Z Wu, et al. Ultralight Mesoporous Magnetic Frameworks by Interfacial Assembly of Prussian Blue Nanocubes[J]. Angew. Chem. Int. Ed., 2014, 53 (11): 2888-2892.

[274] L Hu, P Zhang, H Zhong, et al. Foamlike Porous Spinel $Mn_xCo_{3-x}O_4$ Material Derived from $Mn_3[Co(CN)_6]_2 \cdot nH_2O$ Nanocubes: A Highly Efficient Anode Material for Lithium Batteries[J]. Chem. Eur. J., 2012, 18 (47): 15049-15056.

[275] J Sun, Q Xu. Functional Materials Derived from Open Framework Templates/Precursors: Synthesis and Applications[J]. Energy Environ. Sci., 2014, 7 (7): 2071-2100.

[276] L Zhang, H B Wu, S Madhavi, et al. Formation of Fe_2O_3 Microboxes with Hierarchical Shell Structures from Metal-Organic Frameworks and Their Lithium Storage Properties[J]. J. Am. Chem. Soc., 2012, 134 (42):

17388-17391.

[277] L Hu, N Yan, Q W Chen, et al. Fabrication Based on the Kirkendall Effect of Co_3O_4 Porous Nanocages with Extraordinarily High Capacity for Lithium Storage[J]. Chem. Eur. J., 2012, 18 (29): 8971-8977.

[278] L Hu, P Zhang, Y K Sun, et al. ZnO/Co_3O_4 Porous Nanocomposites Derived from Mofs: Room-Temperature Ferromagnetism and High Catalytic Oxidation of CO[J]. Chemphyschem, 2013, 14(17): 3953-3959.

[279] Y Chen, S Cheng, H Xia, et al. Morphology Controllable Preparation of Nanocube $Mn_3[Fe(CN)_6]_2 \cdot nH_2O$ Particles[J]. Colloid Surf. A-Physicochem. Eng. Asp., 2013, 436: 1140-1144.

[280] F Sousa, M Sousa, I Oliveira, et al. Evaluation of A Low-Cost Adsorbent for Removal of toxic Metal Ions from Wastewater of an Electroplating Factory[J]. J. Environ. Manage., 2009, 90 (11): 3340-3344.

[281] M Regi, F Ballas, D Arcos. Mesoporous Materials for Drug Delivery[J]. Angew. Chem. Int. Ed., 2007, 46 (40): 7548-7558.

[282] Y Yan, F Du, X Shen, et al. Porous SnO_2–Fe_2O_3 Nanocubes with Improved Electrochemical Performance for Lithium Ion Batteries[J]. Dalton Trans., 2014, 43 (46): 17544-17550.

[283] W Hu, X Chen, G Wu, et al. Bipolar and Tri-State Unipolar Resistive Switching Behaviors In $Ag/ZnFe_2O_4/Pt$ Memory Devices[J]. Appl. Phys. Lett., 2012, 101 (6): 063501-063504.

[284] M Sadeghi, W Liu, T Zhang, et al. Role of Photoinduced Charge Carrier Separation Distance in Heterogeneous Photocatalysis: Oxidative Degradation of CH_3OH Vapor in Contact with Pt/TiO_2 and Cofumed TiO_2–Fe_2O_3[J]. J. Phys. Chem., 1996, 100 (50): 19466-19474.

[285] Y Liu, W Yao, D Liu, et al. Enhancement of Visible Light Mineralization

Ability and Photocatalytic Activity of BiPO$_4$/BiOI[J]. Appl. Catal. B: Environ., 2015, 163: 547-553.

[286] S Xu, D Feng, W Shangguan. Preparations and Photocatalytic Properties of Visible-Light-Active Zinc Ferrite-Doped TiO$_2$ Photocatalyst[J]. J. Phys. Chem. C, 2009, 113 (6): 2463-2467.

[287] Y Zhang, J Mu. One-Pot Synthesis, Photoluminescence, and Photocatalysis of Ag/ZnO Composites[J]. Colloid Interface Sci., 2007, 309 (2): 478-484.

[288] Y Lai, M Meng, Y Yu. One-Step Synthesis, Characterizations and Mechanistic Study of Nanosheets-Constructed Fluffy ZnO and Ag/ZnO Spheres Used for Rhodamine B Photodegradation[J]. Appl. Catal. B: Environ., 2010, 100 (3-4): 491-501.

[289] W Lu, G Liu, S Gao, et al. Tyrosine-assisted Preparation of Ag/ZnO Nanocomposites with Enhanced Photocatalytic Performance and Synergistic Antibacterial Activities[J]. Nanotechnology, 2008, 19 (44): 445711.

[290] A Wood, M Giersig, P Mulvaney. Fermi Level Equilibration in Quantum Dot-Metal Nanojunctions[J]. J. Phys. Chem. B, 2001, 105 (37): 8810-8815.

[291] C Cheng, A Amini, C Zhu, et al. Enhanced Photocatalytic Performance of TiO$_2$-ZnO Hybrid Nanostructures[J]. Sci. Rep., 2014, 4 (8): 4181.

[292] M Jakob, H Levanon, P Kamat. Charge Distribution Between Uv-Irradiated TiO$_2$ and Gold Nanoparticles: Determination of Shift in the Fermi Level[J]. Nano Lett., 2003, 3 (3): 353-358.

[293] H Yoo, C Bae, Y Yang, et al. Spatial Charge Separation in Asymmetric Structure of Au Nanoparticle on TiO$_2$ Nanotube by Light-Induced Surface Potential Imaging[J]. Nano Let., 2014, 14 (8): 4413-4417.

[294] T Hirakawa, P Kamat. Charge Separation and Catalytic Activity of Ag@TiO$_2$ Core-Shell Composite Clusters Under Uv-Irradiation[J]. J. Am. Chem. Soc., 2005, 127(11): 3928-3934.

[295] A Sclafani, J M Herrmann. Influence of Metallic Silver and of Platinum-Silver Bimetallic Deposits On the Photocatalytic Activity of Titania (Anatase and Rutile) in Organic and Aqueous Media[J]. J. Photochem. Photobiol. A, 1998, 113(2): 181-188.

[296] X F You, F Chen, J L Zhang, et al. A Novel Deposition Precipitation Method for Preparation of Ag-Loaded Titanium Dioxide[J]. Catal. Lett., 2005, 102(3): 247-250.

[297] H Tahiri, Y A Ichou, J M Herrmann. Photocatalytic Degradation of Chlorobenzoic Isomers in Aqueous Suspensions of Neat and Modified Titania[J]. J. Photochem. Photobiol. A, 1998, 114(3): 219-226.

[298] I M Arabatzis, T Stergiopoulos, D andreeva, et al. Characterization and Photocatalytic Activity of Au/TiO$_2$ Thin Films for Azo-Dye Degradation[J]. J. Catal., 2003, 220(1): 127-135.

[299] H Lv, X Shen, Z Ji, et al. One-Pot Synthesis of Prpo$_4$ Nanorods-Reduced Graphene Oxide Composites and Their Photocatalytic Properties[J]. New J. Chem., 2014, 38(6): 2305-2311.

[300] M S Dresselhaus, I L Thomas. Alternative Energy Technologies[J]. Nature, 2001, 414(6861): 332-337.

[301] J A Turner. Sustainable Hydrogen Production[J]. Science, 2004, 305(5686): 972-974.

[302] S Schuldiner. Hydrogen Overvoltage on Bright Platinum: III. Effect of Hydrogen Pressure[J]. J. Electrochem. Soc., 1959, 106: 891-895.

[303] W C Sheng, H A Gasteiger, Y Shao-Horn. Hydrogen Oxidation and

Evolution Reaction Kinetics on Platinum: Acid Vs Alkaline Electrolytes[J]. J. Electrochem. Soc., 2010, 157 (11): B1529-B1536.

[304] E S Andreiadis, P A Jacques, P D Tran, et al. Molecular Engineering of A Cobalt-Based Electrocatalytic Nanomaterial for H_2 Evolution Under Fully Aqueous Conditions[J]. Nat. Chem., 2013, 5 (1): 48-53.

[305] Q Liu, J Tian, W Cui, et al. Carbon Nanotubes Decorated with CoP Nanocrystals: A Highly Active Non-Noble-Metal Nanohybrid Electrocatalyst for Hydrogen Evolution[J]. Angew. Chem., 2014, 126 (26): 6828-6832.

[306] X Zou, Y Zhang. Noble Metal-Free Hydrogen Evolution Catalysts for Water Splitting[J]. Chem. Soc. Rev., 2015, 44 (5): 5148-5180.

[307] W F Chen, K Sasaki, C Ma, et al. Hydrogen-Evolution Catalysts Based on Non-Noble Metal Nickel- Molybdenum Nitride Nanosheets[J]. Angew. Chem. Int. Ed., 2012, 51 (25): 6131-6135.

[308] J Kibsgaard, T F Jaramillo, F Besenbacher. Building an Appropriate Active-Site Motif into Ahydrogen-Evolution Catalyst with Thiomolybdate $[Mo_3S_{13}]^{2-}$ Clusters[J]. Nat. Chem., 2014, 6 (3): 248-253.

[309] C Wan, Y N Regmi, B M Leonard. Multiple Phases of Molybdenum Carbide As Electrocatalysts for the Hydrogen Evolution Reaction[J]. Angew. Chem. Int. Ed., 2014, 53 (25): 6407-6410.

[310] X Zou, X Huang, A Goswami, et al. Cobalt-Embedded Nitrogen-Rich Carbon Nanotubes Efficiently Catalyze Hydrogen Evolution Reaction at All pH Values[J]. Angew. Chem. Int. Ed., 2014, 53 (17): 4372-4376.

[311] D Voiry, H Yamaguchi, J Li, et al. Enhanced Catalytic Activity in Strained Chemically Exfoliated Ws_2 Nanosheets for Hydrogen Evolution[J]. Nat. Mater., 2013, 12 (9): 850-855.

[312] E J Popczun, J R Mckone, C G Read, et al. Nanostructured Nickel

Phosphide as an Electrocatalyst for the Hydrogen Evolution Reaction[J]. J. Am. Chem. Soc. 2013, 135 (25): 9267-9270.

[313] J Wang, G Wang, S Miao, et al. Graphene-Supported Iron-Based Nanoparticles Encapsulated in Nitrogen-Doped Carbon as a Synergistic Catalyst for Hydrogen Evolution and Oxygen Reduction Reactions[J]. Faraday Discuss., 2014, 176: 135-151.

[314] P Y Ge, M D Scanlon, P Peljo, et al. Multichannel HSO_4^- Recognition Promoted by A Bound Cation within Aferrocene-Based Ion Pair Receptor[J]. Chem. Commun., 2012, 48 (54): 6848-6850.

[315] M Tavakkoli, T Kallio, O Reynaud, et al. Single-Shell Carbon-Encapsulated Iron Nanoparticles: Synthesis and High Electrocatalytic Activity for Hydrogen Evolution Reaction[J]. Angew. Chem., 2015, 54 (15): 4535-4538.

[316] Y Ding, P Kopold, K Hahn, et al. Facile Solid-State Growth of 3d Well-Interconnected Nitrogen-Rich Carbon Nanotube-Graphene Hybrid Architectures for Lithium-Sulfur Batteries[J]. Adv. Funct. Mater., 2015, 26 (7): 1681-1693.

[317] X Zheng, J Deng, N Wang, et al. Podlike N-Doped Carbon Nanotubes Encapsulating Feni Alloy Nanoparticles: High-Performance Counter Electrode Materials for Dye-Sensitized Solar Cells[J]. Angew. Chem. Int. Ed., 2014, 53 (27): 7023-7027.

[318] Q Li, H Pan, D Higgins, et al. Metal–Organic Framework-Derived Bamboo-Like Nitrogen-Doped Graphene Tubes as an Active Matrix for Hybrid Oxygen-Reduction Electrocatalysts[J]. Small, 2015, 11 (12): 1443-1452.

[319] Q Li, P Xu, W Gao, et al. Graphene/Graphene-Tube Nanocomposites

Templated from Cage-Containing Metal-Organic Frameworks for Oxygen Reduction in Li–O_2 Batteries[J]. Adv. Mater., 2014, 26（9）: 1378-1386.

[320] X Wang, Q Weng, X Liu, et al. Atomistic Origins of High Rate Capability and Capacity of N-Doped Graphene for Lithium Storage[J]. Nano Lett., 2014, 14（3）: 1164-1171.

[321] Q Li, P Xu, B Zhang, et al. One-Step Synthesis of Mn_3O_4/Reduced Graphene Oxide Nanocomposites for Oxygen Reduction in Nonaqueous Li–O_2 Batteries[J]. Chem. Commun., 2013, 49（92）: 10838-10840.

[322] Q G He, Q Li, S Khene, et al. High-Loading Cobalt Oxide Coupled with Nitrogen-Doped Graphene for Oxygen Reduction in Anion-Exchange-Membrane Alkaline Fuel Cells[J]. J. Phys. Chem. C, 2013, 117（17）: 8697-8707.

[323] G Wu, N H Mack, W Gao, et al. Nitrogen-Doped Graphene-Rich Catalysts Derived from Heteroatom Polymers for Oxygen Reduction in Nonaqueous Lithium–O_2 Battery Cathodes[J]. ACS Nano, 2012, 6（11）: 9764-9776.

[324] 夏丹葵, 戴永年. 纯锌蒸发规律研究[J]. 有色金属, 1992, 44（4）: 52-55.

[325] T Sharifi, G Hu, X Jia, et al. Formation of Active Sites for Oxygen Reduction Reactions by Transformation of Nitrogen Functionalities in Nitrogen-Doped Carbon Nanotubes[J]. Acs Nano, 2012, 6（10）: 8904-8912.

[326] Y Liu, X Xu, P Sun, et al. N-Doped Porous Carbon Nanosheets with Embedded Iron Carbide Nanoparticles for Oxygen Reduction Reaction in Acidic Media[J]. Int. J. Hydrog. Energ., 2015, 40（13）, 4531-4539.

[327] C Mao, A Kong, Y Wang, et al. MIL-100 Derived Nitrogen-Embodied Carbon Shells Embedded with Iron Nanoparticles[J]. Nanoscale, 2015, 7

（24）：10817-10822.

[328] F Zheng, Y Yang, Q Chen. High Lithium Anodic Performance of Highly Nitrogen-Doped Porous Carbon Prepared from a Metal-Organic Framework[J]. Nat. Commun., 2014, 5（5）：5261.

[329] C N. Rao, A K Sood, K S Subrahmanyam, et al. Graphene: the New Two-Dimensional Nanomaterial[J]. Angew. Chem. Int. Ed., 2009, 48(42): 7752-7777.

[330] M S Dresselhaus, A Jorio, M Hofmann, et al. Perspectives On Carbon Nanotubes and Graphene Raman Spectroscopy[J]. Nano Lett., 2010, 10（3）：751-758.

[331] M M Lucchese, F Satavle, E H Martins Ferreira, et al. Quantifying ion-induced defects and Raman relaxation length in graphene[J]. Carbon, 2010, 48（5）：1592-1597.

[332] H Zhang, H Li, X Li, et al. Pyrolyzing Cobalt Diethylenetriamine Chelate on Carbon (CoDETA/C) as a Family of Non-Precious Metal Oxygen Reduction Catalyst[J]. Int J Hydrogen Energy, 2014, 39（1）：267-276.

[333] Z Liu, G Zhang, Z Lu, et al. One-Step Scalable Preparation of N-Doped Nanoporous Carbon as a Highperformance Electrocatalyst for the Oxygen Reduction Reaction[J]. Nano Res, 2013, 6（4）：293-301.

[334] Q Liu, J Tian, W Cui, et al. Carbon Nanotubes Decorated with Cop Nanocrystals: A Highly Active Non-Noble-Metal Nanohybrid Electrocatalyst for Hydrogen Evolution[J]. Angew Chem Int Ed Engl. 2014, 53（26）：6710-6714.

[335] Z P Huang, Z B Chen, Z Z Chen, et al. $Ni_{12}P_5$ Nanoparticles as an Efficient Catalyst for Hydrogen Generation Via Electrolysis and Photoelectrolysis[J]. ACS Nano, 2014, 8（8）：8121-8129.

[336] H Lv, Z Xi, Z Chen, et al. A New Core/Shell Niau/Au Nanoparticle Catalyst with Pt-Like Activity for Hydrogen Evolution Reaction[J]. J. Am. Chem. Soc., 2015, 137 (18): 5859-5862.

[337] S H Ahn, S Hwang, S Yoo, et al. Electrodeposited Ni Dendrites with High Activity and Durability for Hydrogen Evolution Reaction In Alkaline Water Electrolysis[J]. J. Mater. Chem., 2012, 22 (30): 15153-15159.

[338] 宋全生, 唐致远, 郭鹤桐. 电沉积镍-储氢合金复合电极的析氢电催化性能[J]. 天津大学学报(自然科学与工程技术版), 2005, 38 (6): 508-512.

[339] S Badrayyana, D K Bhat, S Shenoy, et al. Novel Feeni-Graphene Composite Electrode for Hydrogen Production[J]. Int. J. Hydrog. Energ., 2015, 40 (33): 10453-10462.

[340] B E Conway, G Jerkiewicz. Relation of Energies and Coverages of Underpotential and Overpotential Deposited H at Pt and Other Metals to the 'Volcano Curve' for Cathodic H_2 Evolution Kinetics[J]. Electrochim. Acta, 2000, 45 (25-26): 4075-4083.

[341] C He, N Zhao, C Shi, et al. Fabrication of Nanocarbon Composites Using in Situ Chemical Vapor Deposition and Their Applications[J]. Adv. Mater., 2015, 27 (36): 5422-5431.

[342] C Seah, S Chai, A R Mohamed. Mechanisms of Graphene Growth by Chemical Vapour Deposition on Transition Metals[J]. Carbon, 2014, 70 (4): 1-21.

[343] H An, W Lee, J Jung. Graphene Synthesis on Fe Foil Using Thermal CVD[J]. Curr Appl Phys, 2011, 11 (S4): S81-S85.

[344] C Mattevi, H Kim, M Chhowalla. A Review of Chemical Vapour

Deposition of Graphene on Copper[J]. J Mater Chem, 2011, 21 (10): 3324-3334.

[345] X Zheng, J Deng, N Wang, et al. Podlike N-Doped Carbon Nanotubes Encapsulating Feni Alloy Nanoparticles: High-Performance Counter Electrode Materials for Dye-Sensitized Solar Cells[J]. Angew. Chem. Int. Ed., 2014, 53 (27): 7023-7027.

[346] I Martin-Gullon, J Vera, J A Conesa, et al. Differences Between Carbon Nanofibers Produced Using Fe and Ni Catalysts in a Floating Catalyst Reactor[J]. Carbon, 2006, 44 (8): 1572-1580.

[347] X H Li, W C H Choy, X G Ren, et al. Highly Intensified Surface Enhanced Raman Scattering by Using Monolayer Graphene as the Nanospacer of Metal Film–Metal Nanoparticle Coupling System[J]. Adv. Funct. Mater., 2014, 24 (21): 3114-3122.

[348] D Graf, F Molitor, K Ensslin, et al. Spatially Resolved Raman Spectroscopy of Single- and Few-Layer Graphene[J]. Nano Lett., 2007, 7 (2): 238-242.

[349] A Barras, M R Das, R R Devarapalli, et al. One-Pot Synthesis of Gold Nanoparticle/Molybdenum Cluster/Graphene Oxide Nanocomposite and Its Photocatalytic Activity[J]. Appl. Catal. B, 2013, S130-131 (6): 270-276.

[350] J Chang, M H Jin, F Yao, et al. Asymmetric Supercapacitors Based on Graphene/MnO_2 Nanospheres and Graphene/MoO_3 Nanosheets with High Energy Density[J]. Adv. Funct. Mater., 2013, 23 (40): 5074-5083.

[351] X S Qi, Y Deng, W Zhong, et al. Controllable and Large-Scale Synthesis of Carbon Nanofibers, Bamboo-Like Nanotubes, and Chains of Nanospheres Over Fe/SnO_2 and Their Microwave-Absorption

Properties[J]. J. Phys. Chem. C, 2010, 114 (2): 808-814.

[352] C Wang, Z Y Guo, W Shen, et al. B-Doped Carbon Coating Improves the Electrochemical Performance of Electrode Materials for Li-Ion Batteries[J]. Adv. Funct. Mater., 2014, 24 (35): 5511-5521.

[353] K T Lee, J C Lytle, N S Ergang, et al. Synthesis and Rate Performance of Monolithic Macroporous Carbon Electrodes for Lithium-Ion Secondary Batteries[J]. Adv. Funct. Mater., 2005, 15 (15): 547-556.

[354] M Ohring. The Materials Science of Thin Films[M]. London: Academic Press, 1991.

[355] Z H Wen, S Q Ci, F Zhang, et al. Nitrogen-Enriched Core-Shell Structured Fe/Fe$_3$C-C Nanorods as Advanced Electrocatalysts for Oxygen Reduction Reaction[J]. Adv. Mater., 2012, 24 (11): 1399-1404.

[356] J A Rodríguez-Manzo, C Pham-Huu, F Banhart. Graphene Growth by A Metal-Catalyzed Solid-State Transformation of Amorphous[J]. Carbon, 2011, 5 (2): 1529-1534.

[357] N M Rodriguez, M S Kim, F Fortin, et al. Carbon Deposition on Iron-Nickel Alloy Particles[J]. Appl. Catal. A: Gen., 1997, 148 (2): 265-282.

[358] C Park, R T K Baker. Carbon Deposition on Iron-Nickel During Interaction with Ethylene-Hydrogen Mixtures[J]. J. Catal., 1998, 179 (2): 361-374.

[359] A Tanaka, S Yoon, I Mochida. Preparation of Highly Crystalline Nanofibers on Fe and Fe-Ni Catalysts with a Variety of Graphene Plane Alignments[J]. Carbon, 2004, 42 (3): 591-597.

[360] N Park, D Lee, I Ko, et al. Rapid Consolidation of Nanocrystalline Al$_2$O$_3$ Reinforced Ni-Fe Composite from Mechanically Alloyed Powders by

High Frequency Induction Heated Sintering[J]. Ceram. Int., 2009, 35 (8): 3147-3151.

[361] W F Chen, J T Mucherman, E Fujita. Recent Developments in Transition Metal Carbides and Nitrides as Hydrogen Evolution Electrocatalysts[J]. Chem. Commun., 2013, 49 (79): 8896-8909.

[362] D S Kong, J J Cha, H T Wang, et al. First-Row Transition Metal Dichalcogenide Catalysts for Hydrogen Evolution Reaction[J]. Energy Environ. Sci., 2013, 6 (12): 3553-3558.

[363] J O M Bockris, E C Potter. The Mechanism of the Cathodic Hydrogen Evolution Reaction[J]. J. Electrochem. Soc., 1952, 99: 169-186.

[364] Z P Huang, Z B Chen, Z Z Chen, et al. $Ni_{12}P_5$ Nanoparticles as an Efficient Catalyst for Hydrogen Generation Via Electrolysis and Photoelectrolysis[J]. ACS Nano, 2014, 8 (8): 8121-8129.

[365] H Zhang, Z Ma, J Duan, et al. Active Sites Implanted Carbon Cages in Core-Shell Architecture: Highly Active and Durable Electrocatalyst for Hydrogen Evolution Reaction[J]. ACS Nano, 2016, 10 (1): 684-694.

[366] A Patru, P Antitomaso, R Sellin, et al. Size and Strain Dependent Activity of Ni Nano and Micro Particles for Hydrogen Evolution Reaction[J]. Int. J. Hydrogen Energy., 2013, 38 (27): 11695-11708.

[367] J Lu, W Zhou, L Wang, et al. Core-Shell Nanocomposites Based on Gold Nanoparticle@Zinc-Iron-Embedded Porous Carbons Derived from Metal-Organic Frameworks as Efficient Dual Catalysts for Oxygen Reduction and Hydrogen Evolution Reactions[J]. ACS Catal., 2016, 6 (2): 1045-1053.

[368] X Mao, J Kwon, E K Koh, et al. Ligand Exchange Procedure for Bimetallic Magnetic Iron-Nickel Nanocrystals toward Biocompatible

Activities[J]. ACS Appl Mater Interfaces, 2015, 7 (28): 15522-15530.

[369] Q Wei, X Tong, G Zhang, et al. Nitrogen-Doped Carbon Nanotube and Graphene Materials for Oxygen Reduction Reactions[J]. Catalysts, 2015, 5 (3): 1574-1602.

[370] C Mao, A Kong, Y Wang, et al. MIL-100 Derived Nitrogen-Embodied Carbon Shells Embedded with Iron Nanoparticles[J]. Nanoscale, 2015, 7 (24): 10817-10822.

[371] B Hinnemann, P G Moses, J Bonde, et al. Biomimetic Hydrogen Evolution: MoS_2 Nanoparticles as Catalyst for Hydrogen Evolution[J]. J. Am. Chem. Soc., 2005, 127 (15): 5308-5309.

[372] J K Nørskov, T Bligaard, A Logadottir, et al. Trends in the Exchange Current for Hydrogen Evolution[J]. J. Electrochem. Soc., 2005, 152(3): J23-J26.

[373] D H Deng, L Yu, X Q Chen, et al. Iron Encapsulated within Pod-Like Carbon Nanotubes for Oxygen Reduction Reaction[J]. Angew. Chem. Int. Ed., 2013, 52 (1): 371-375.

[374] S Simonetti, C Canto. A Model for Atomic Hydrogen-Bimetal Interactions[J]. Int. J. Hydrogen Energy, 2012, 37 (19): 14730-14734.

[375] W Yang, X Liu, X Yue, et al. Bamboo-Like Carbon Nanotube/Fe_3c Nanoparticle Hybrids and Their Highly Efficient Catalysis for Oxygen Reduction[J]. J. Am. Chem. Soc., 2015, 137 (4): 1436-1439.

[376] L Liao, S N Wang, J J Xiao, et al. A Nanoporous Molybdenum Carbide Nanowire As an Electrocatalyst for Hydrogen Evolution Reaction[J]. Energy Environ. Sci., 2014, 7 (1): 387-392.

[377] Y D Ma, Y Dai, M Guo, et al. Graphene Adhesion on MoS_2 Monolayer: an Ab Initio Study[J]. Nanoscale, 2011, 3 (9): 3883-3887.

[378] H Vrubel, X Hu. Molybdenum Boride and Carbide Catalyze Hydrogen Evolution in Both Acidic and Basic Solutions[J]. Angew. Chem. Int. Ed., 2012, 124(51): 12875-12878.